Embedded Digital Control with Microcontrollers

Embedded Digital Control with Microcontrollers

Implementation with C and Python

Cem Ünsalan
Marmara University

Duygun E. Barkana
Yeditepe University

H. Deniz Gürhan
Yeditepe University

Published by John Wiley & Sons, Inc., Hoboken, New Jersey. All rights reserved.
Published simultaneously in Canada.

For general information on our other products and services or for technical support, please contact our Customer Care Department within the United States at (800) 762-2974, outside the United States at (317) 572-3993 or fax (317) 572-4002.

Wiley also publishes its books in a variety of electronic formats. Some content that appears in print may not be available in electronic formats. For more information about Wiley products, visit our web site at www.wiley.com.

Library of Congress Cataloging-in-Publication Data applied for:

ISBN: 9781119576525

Cover design by Wiley
Cover image: businessman/depositphotos

Set in 9.5/12.5pt STIXTwoText by SPi Global, Chennai, India

To our families.

Contents

Preface

We are surrounded by systems performing specific tasks for us. There are also control systems designed to improve existing system characteristics. To do so, an input signal (possibly originating from a sensor) is acquired. The control system generates a control signal for this input. Hence, the desired system output is obtained.

The designed control system can be either in analog or digital form. Analog control systems are constructed by either electrical or mechanical elements. With the arrival of embedded systems, digital control became the new standard. Recent microcontrollers provide a cheap and powerful platform for this purpose. This book aims to introduce implementation methods and theory of digital control systems on microcontrollers via focusing on real-life issues.

Python, MicroPython (the modified form of Python to be used in embedded systems), and C will serve as the programming languages throughout the book. Python will be extensively used in explaining theoretical digital control concepts. MicroPython and C languages will be the main mediums for microcontroller based implementation. Hence, the reader will develop and implement a digital controller for a given problem.

We took an undergraduate engineering student and hobbyist as benchmark in explaining digital control concepts. Therefore, a professional engineer may also benefit from the book. We pick the STM32 board with an Arm Cortex-M microcontroller on it. Hence, the reader may find a wide variety of applications besides the ones considered in this book. As a result, we expect the reader to become familiar with the basic and advanced digital control concepts in action.

Istanbul, Turkey
June 2020

Cem Ünsalan
Duygun E. Barkana
H. Deniz Gürhan

About the Companion Website

This book is accompanied by a companion website:

**www.wiley.com/go/Unsalan/Embedded_Digital_Control_with_
Microcontrollers**

The website includes:
1. C and Python codes and libraries used in the book.
2. C and Python codes and supplementary material for the end of chapter applications.
3. Images used in the book.
4. PowerPoint slides for the instructors.
5. Solution manual for the end of chapter questions **(only to the instructors who adopted the book)**.

1

Introduction

This book aims to introduce digital control systems via practical applications. Therefore, we will briefly introduce the system and control theory concepts in this chapter. Then, we will emphasize how this book differs from the ones in literature. Besides, we will summarize the concepts to be explored in the book. Hence, the reader will have necessary background for the following chapters.

1.1 What is a System?

A system can be defined as the combination of parts to carry out a specific task. Let us pick the Pololu Zumo robot (https://www.pololu.com/product/2510/) in Figure 1.1. This is a system composed of four main parts as chassis, motors, tracks with sprockets, and battery. When energy is fed to the motors, they move the chassis via rotating sprockets.

We can add two more modules to the Zumo robot as a control unit (such as microcontroller) and reflectance sensor. These modules can be used to add autonomy to the robot such that it can follow a line drawn on ground. To do so, we will need a "control" action. Let us introduce it next.

1.2 What is a Control System?

Control is the act of producing a desired output for a given input. The control system is used for this purpose. For our Zumo robot, our aim is following the line. Hence, the robot system should be guided by a control system to follow the line. To be more specific, the control system should get the reference input signal (as position of the line) and current position of the robot from the reflectance sensor; form an error signal by their difference; generate necessary control signals to

Embedded Digital Control with Microcontrollers: Implementation with C and Python,
First Edition. Cem Ünsalan, Duygun E. Barkana, and H. Deniz Gürhan.
© 2021 The Institute of Electrical and Electronics Engineers, Inc. Published 2021 by John Wiley & Sons, Inc.
Companion website: www.wiley.com/go/Unsalan/Embedded_Digital_Control_with_Microcontrollers

Figure 1.1 PololuZumo robot. (Source: Pololu Robotics & Electronics, pololu.com.)

the motors (on the chassis); and guide the robot (system) accordingly. Yet another example is controlling temperature inside the refrigerator. Here, refrigerator is the system. The desired temperature value is set by the user as the reference signal. The actual temperature value inside the refrigerator is measured by a sensor. The control system sets the internal temperature to the desired value by using a cooler. As can be seen in both examples, the control system is used to generate a desired output for a given input.

The control system may be classified either as analog or digital depending on its construction. If the system is only formed by analog components, then it is analog. Early control systems were of this type. As the microcontroller and embedded systems are introduced, digital controllers became dominant. The main reason for this shift is that the control system developed in digital systems is a code snippet which can be modified easily. Therefore, this book aims to introduce digital control methods implemented on microcontrollers. We will form a general setup for this purpose as in Figure 1.2.

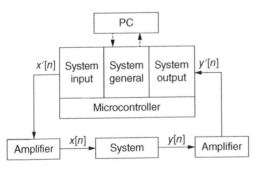

Figure 1.2 General setup for a digital control system.

There is a system to be controlled in Figure 1.2. The microcontroller is the medium digital control algorithms are implemented on. The generated control signal, $x[n]$, is fed to the system as input by an amplifier. Likewise, system output signal, $y[n]$, is fed to the microcontroller for further processing. We can also connect the microcontroller to PC to send or receive data. We will use the setup in Figure 1.2 for almost all control operations throughout the book.

1.3 About the Book

There are several good books on digital control. We can group them into two categories. The first category consists of books on theoretical concepts in discrete-time control systems (Xue et al. 2007; Dorf and Bishop 2010; Burns 2001; Chen 2006; Corke 2017; Franklin et al. 2006; Ghosh 2004; Gopal 2003; Golnaraghi and Kuo 2010; Mandal 2010; Moudgalya 2008; Goodwin et al. 2000; Tewari 2002; Ogata 1995; Phillips et al. 2015; Starr 2006; Wescott 2006). These are useful in understanding theoretical foundations of digital control. Some books in this category also provide MATLAB-based implementation (Xue et al. 2007; Dorf and Bishop 2010; Chen 2006; Corke 2017; Mandal 2010; Tewari 2002). The reader can consult these in case theory is not sufficient.

The second category consists of books on practical aspects and implementation details of digital control systems on microcontrollers. Unfortunately, there are few books in this category (Braunl 2006; Forrai 2013; Ibrahim 2006; Ledin 2004; Hristu-Varsakelis and Levine 2005). Besides, there is no book on digital control applications with the Python programming language. This book aims to fill these gaps. Hence, it handles theoretical digital control concepts by Python. Besides, we benefit from MATLAB in controller design and system identification steps. Then, digital control concepts are implemented and realized on a low-level microcontroller using MicroPython and C languages. Via this approach, we aim to bridge the gap between theory and practice.

The book is composed of 12 chapters. We devote Chapter 2 to introduce hardware to be used in the book. Details of software platforms to be used are given in Chapter 3. Basic digital signal processing and control concepts are provided in Chapter 4. We lay the framework for modeling the continuous-time system to be controlled in Chapters 5 and 6. We devote Chapters 7 and 8 to transfer function-based control system analysis and design techniques. We introduce state-space analysis and design in Chapters 9 and 10. We provide adaptive control methods in Chapter 11. Finally, Chapter 12 introduces advanced methods and practical digital control applications. Therefore, this final chapter aims to show the reader how the concepts introduced in the book can be implemented to solve actual real-life problems.

The reader can reach all C and Python codes introduced throughout the book in the accompanying book website. The complete project setup for the end of chapter applications is also available in the same website. The instructors adopting the book for their course can reach the solution manual for the end of chapter problems and projects introduced in Chapter 12 from the publisher.

2

Hardware to be Used in the Book

Throughout the book, we will not only introduce digital control concepts from a theoretical perspective, we will also implement them on embedded hardware using C and Python languages. Therefore, the reader should become familiar with the hardware to be used. This chapter aims to introduce these concepts. To do so, we will assume a novice user as our target. Besides, we will cover all hardware topics as abstract as possible. Hence, they can give insight on similar platforms. As for embedded hardware, we will pick the STMicroelectronics NUCLEO-F767ZI development board (STM32 board) and STM32F767ZI microcontroller (STM32 microcontroller) on it. These are the mediums our C and Python codes for digital control will be implemented on. Afterward, we will introduce the DC motor, its driver, and related hardware to be used in examples throughout the book. Finally, we will introduce other systems and sensors which can be used in advanced applications. As all the hardware is introduced, we will be ready to use them in practical digital control applications in solving real-life problems.

2.1 The STM32 Board

Our C and Python codes for digital control will run on the STM32F767ZI microcontroller. However, we cannot use this microcontroller alone since it needs extra hardware to operate. There should be programming and debugging circuitry accompanying the microcontroller. Hence, it can be programmed easily. For these reasons, development boards emerged. These have all the necessary circuitry on them. Therefore, they provide a complete environment to use the microcontroller. In this book, we pick the STMicroelectronics NUCLEO-F767ZI development board for this purpose. For the sake of brevity, we will call it as the STM32 board from this point on. In this section, we will provide general information about the board. We will also provide the pin layout. This will be mandatory to interface the

Embedded Digital Control with Microcontrollers: Implementation with C and Python,
First Edition. Cem Ünsalan, Duygun E. Barkana, and H. Deniz Gürhan.
© 2021 The Institute of Electrical and Electronics Engineers, Inc. Published 2021 by John Wiley & Sons, Inc.
Companion website: www.wiley.com/go/Unsalan/Embedded_Digital_Control_with_Microcontrollers

microcontroller with outside world. We will also evaluate methods to program and power the board.

2.1.1 General Information

The STM32 board is as in Figure 2.1. This board has an STM32F767ZI microcontroller on it. Besides, it has three user LEDs, one user push button and one reset push button. It has a 32768 kHz crystal oscillator on it. Board connectors, which lead to access to all microcontroller input and output pins, are compatible with Arduino. The STM32 board has an on-board ST-LINK debugger/programmer with USB connectivity. For more information on the development board, please see https://www.st.com/en/evaluation-tools/nucleo-f767zi.html.

2.1.2 Pin Layout

Pin layout of the STM32 board is as in Figure 2.2. As can be seen in this figure, pins are gathered under four connectors as CN7, CN8, CN9, and CN10. Within each connector, there are input and output pins labeled as PA, PB, PC, PD, PE, PF, PG, and PH. The STM32 board also has two unmounted connecters called CN11 and CN12. We will not explain them here.

Pins on the STM32 board can be used for various purposes. Usage areas of each pin are summarized in Appendix A from this book's perspective. We will explore each property separately in the following sections.

We should also mention the pin connection of onboard green, blue, red LEDs, and push button on the STM32 board. The onboard green LED is connected to

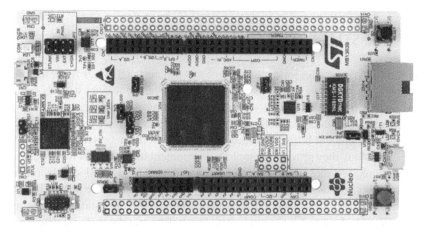

Figure 2.1 The STM32 board. (Source: STMicroelectronics. Used with permission.)

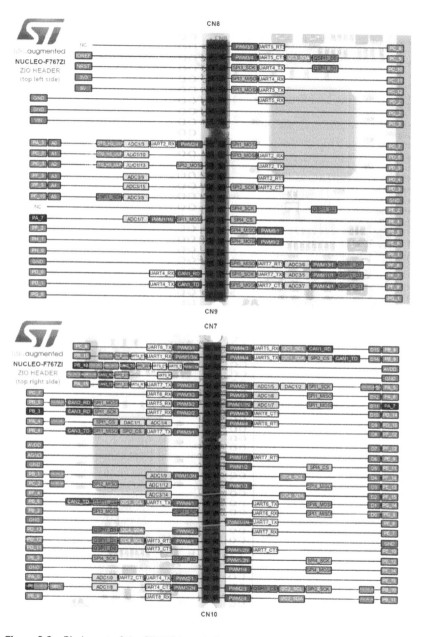

Figure 2.2 Pin layout of the STM32 board. (Source: Nucleo-F767ZI Zio Header, used with permission of STMicroelectronics.)

pin PB0. The onboard blue LED is connected to pin PB7. The onboard red LED is connected to pin PB14. The onboard user push button is connected to pin PC13. The reader should use the mentioned pins to reach the onboard LEDs and push button in the following chapters.

2.1.3 Powering and Programming the Board

The microcontroller on the STM32 board can be programmed easily by the on-board ST-LINK debugger/programmer. To do so, we should connect the board to PC via USB connection. We will introduce methods to program the microcontroller using this connection in Chapter 3.

The USB connection for debugging/programming purposes can also be used to power the board. Hence, whenever the board is connected to PC, it runs by the provided power. Likewise, we can use a USB battery pack to power the board in the same setup. We can also use an external power supply to power the board. To do so, we should use the relevant pins on the board. These will be sufficient to use the board in stand-alone applications.

The STM32 microcontroller operates within the voltage range of 1.71–3.6 V. We call this value as supply voltage (V_{DD}) throughout the book. Let us explain the voltage range in detail. The actual working voltage level for the microcontroller is 3.6 V. This supply voltage may be decreased till 1.7 V for some low power operations. To note here, some peripheral modules will not work at this voltage level.

2.2 The STM32 Microcontroller

As mentioned earlier, the STM32 board is equipped with the STM32F767ZI microcontroller. For the sake of brevity, we will call it as the STM32 microcontroller from this point on. We will evaluate the properties of this microcontroller in this section. Let us start with the functional block diagram of the STM32 microcontroller given in Figure 2.3. As can be seen in this figure, there are several modules in the microcontroller. Besides, interconnection of these modules is complex. However, we will only cover the relevant modules to be used throughout the book here. For more information on other modules of the STM32 microcontroller, please see https://www.st.com/en/microcontrollers-microprocessors/stm32f767zi.html.

2.2.1 Central Processing Unit

Central processing unit (CPU) is the main module responsible for organizing all operations within the microcontroller. This is done by executing the code fed to it. The code can be written in C or MicroPython languages (for this book) with

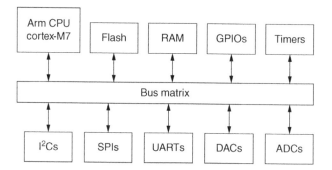

Figure 2.3 Functional block diagram of the STM32 microcontroller.

different execution phases. For more information on this topic, please consult a microcontroller book such as (Yiu 2013).

The CPU in the STM32 microcontroller is based on the Arm Cortex-M7 architecture. Let us explain this in more detail. Arm produces CPU cores in soft form (called as IPs). Microcontroller vendors, such as ST Microelectronics, purchase the right to use these IPs and develop microcontroller hardware. The advantage of this model is as follows. When different vendors use the same CPU core by Arm, they will have the same instruction set and properties. Therefore, the code developed for one microcontroller can be ported to another microcontroller from a different vendor. There is one important issue here. The microcontroller is not only composed of CPU. It also has peripheral units (to be explained next). These may differ for different vendors. Therefore, it may not be possible to directly port the code generated for one microcontroller to another (produced by a different vendor) when peripheral units are used.

Operations within the CPU are done in clock cycles. Before going further, let us first explain what the clock signal means. Clock is a periodic square wave generated by an oscillator. Frequency of the clock signal is measured in Hertz (Hz) which indicates how many periodic pulses occur in one second. The CPU depends on the clock signal. For the STM32 microcontroller, this clock frequency is maximum 216 MHz. The processor performs an action corresponding to an instruction execution phase with each clock cycle. Assuming that an instruction requires four clock cycles to execute, the CPU can process 54000000 instructions per second. Hence, higher the frequency of clock signal, the faster operations are performed within the CPU.

2.2.2 Memory

The microcontroller needs a medium to keep the code to be executed and variables to be operated on. The relevant medium in the microcontroller is called memory.

Unless the microcontroller is using an additional external memory, the core memory is always on the microcontroller chip.

There are two memory regions on the microcontroller as flash and RAM. Codes to be executed are kept in the flash. As power of the microcontroller is turned down, codes remain there. Therefore, flash resembles the solid-state drive (SSD) on PC. Although the recent SSD storage size for a PC is reasonable, memory space in flash of a microcontroller is very limited. For the STM32 microcontroller, this is 2 MB. Therefore, the user should prepare his or her digital control code such that it does not exceed this limit. Fortunately, most digital control algorithms fit into this space.

The medium for temporary storage in the microcontroller is called RAM. Hence, variables to be executed in the code are kept there. This is similar to the RAM on PC with one difference. The RAM on the microcontroller is very limited in storage size. For the STM32 microcontroller, the RAM size is 512 kB. Therefore, the reader should use this space with care.

2.2.3 Input and Output Ports

A port in the microcontroller mean is a group of pins (or wires). These are used to input data to the microcontroller or output data from the microcontroller. Hence, the microcontroller can interact with the outside world through its input and output ports. Here, the processed data can be analog or digital. The STM32 microcontroller has 114 pins (arranged in eight ports called A, B, C, D, E, F, G, and H). All these pins can be used as input or output. They can also be used for other operations as well. Therefore, they are called general purpose input and output (GPIO). We summarized the usage area of each pin in Appendix A.

Digital input and output values are processed in voltage levels as 0 V and V_{DD} (supply voltage). Within the microcontroller code, these correspond to logic level zero and one, respectively. Therefore, the reader should always remember that when the logic level one is fed to output from a pin of the microcontroller, the voltage there is V_{DD}. Similarly, when the logic level zero is fed to output from a pin of the microcontroller, the voltage there is 0 V.

2.2.4 Timer Modules

The timer module is responsible for all time-based operations within the microcontroller. The timer can be taken as a simple counter fed by a clock signal. Based on frequency of the used clock signal and the maximum (or minimum) count value, actual time-based operations can be done within the microcontroller. Since these operations are important, the STM32 microcontroller has 18 timer modules. Ten of these modules are for general purpose; two of them are for advanced control operations; two of them are basic timers; one of them is low power timer; one of them is the Systick timer; and two of them are watchdog timers. Although the microcontroller has such a diverse set of timer modules, we will use general purpose timer modules most of the times. For more detail on the usage of other timer modules, please see https://www.st.com/en/microcontrollers-microprocessors/stm32f767zi.html.

2.2.5 ADC and DAC Modules

The STM32 microcontroller can process analog voltages besides digital ones. To do so, there are specific pins which can accept analog voltage as input. Likewise, there are specific pins which can feed analog voltage to output. These pins are tabulated in Appendix A.

If the user wants to input analog voltage, analog to digital converter (ADC) module of the microcontroller should be used. This module converts a given analog voltage to digital form. We will provide the theory behind this operation in Chapter 5. We will use the ADC module when a sensor with analog input is connected to the microcontroller. The STM32 microcontroller has three ADC modules with 12-bit output. Each module converts a given analog voltage to 12 bits in digital form. Hence, the converted value can be processed within the microcontroller.

Digital data within the microcontroller can be fed to output from an appropriate pin by the digital to analog converter (DAC) module of the microcontroller. We can think of this module as the complement of ADC. We will provide the theory behind the DAC operation in Chapter 5. The STM32 microcontroller has two 12-bit DAC modules which convert a given 12 bit digital data to analog voltage. Although the STM32 microcontroller has a dedicated DAC module, some microcontrollers lack it. They benefit from the pulse width modulation (PWM) method to generate approximate analog voltage by varying width of a square wave. For more information on how PWM can be generated within the STM32 microcontroller, please see Chapter 3.

2.2.6 Digital Communication Modules

The microcontroller may need to communicate with external devices such as sensor modules or other microcontrollers for some applications. There are dedicated digital communication modules within the microcontroller for this purpose. These have specific communication modes such as universal asynchronous receiver/transmitter (UART), universal synchronous/asynchronous receiver/transmitter (USART), serial peripheral interface (SPI), inter integrated circuit (I^2C), and controller area network (CAN).

The STM32 microcontroller has four UART, four USART, six SPI, four I^{2}C, and three CAN modules. These modules have dedicated pins as explained in Appendix A. We will explain the usage of these modules in the following chapters whenever needed. For more detail on these modules, please see https://www.st .com/en/microcontrollers-microprocessors/stm32f767zi.html.

2.3   System and Sensors to be Used Throughout the Book

Besides the STM32 board, we will be using actual system and sensors to explain digital control concepts throughout the book. Therefore, we pick the Pololu 75:1 metal gearmotor with encoder as the actual system to be controlled. In order to use this motor, we need a motor driver module. Therefore, we pick the X-NUCLEO-IHM04A1 dual brush DC motor drive expansion board from STMicroelectronics. We also pick the encoder of the motor as the sensor module. Finally, we pick the FT232 UART to USB converter module to send data from the STM32 microcontroller to PC.

2.3.1   The DC Motor

The DC motor is extensively used in digital control applications. Therefore, we pick it as an actual system to be controlled throughout the examples in the book. There are several motors having different properties in the market. We specifically selected the Pololu metal gearmotor with encoder having the exact name "Pololu 75:1 metal gearmotor 25D × 69L mm HP 12 V with 48 CPR encoder." This motor is well documented and the reader can purchase it easily. In this section, we will explain its properties.

2.3.1.1   Properties of the DC Motor

The Pololu gearmotor is cylindrical with diameter around 25 mm. It has a brushed DC motor combined with a 74.83:1 gearbox. This gearbox decreases the rotation speed meanwhile increasing torque. The motor also has an integrated encoder which can be used as a sensor (to be explained in Section 2.3.3). Image of the motor is given in Figure 2.4.

The brushed DC motor requires 12 V to operate. Its maximum speed is 130 rpm. The motor requires 0.3 A current when idle and 5.6 A (maximum) when loaded. More information on the brushed DC motor can be found in https://www.pololu .com/product/4846. For the sake of brevity, we will call it as DC motor from this point on.

Figure 2.4 Pololu DC motor. (Source: Pololu Robotics & Electronics, pololu.com.)

2.3.1.2 Pin Layout

The DC motor has six color-coded pins to power the motor/encoder and feed encoder output. These pins and their functions are given in Table 2.1.

Table 2.1 Pin usage table for the DC motor.

Pin	Color	Function
1	Red	Motor power supply
2	Black	Motor power supply
3	Green	Encoder ground
4	Blue	Encoder supply voltage, V_{CC}, (3.5–20 V)
5	Yellow	Encoder A output
6	White	Encoder B output

2.3.1.3 Power Settings

We know that the DC motor requires 5.6 A current when fully loaded. However, supplying this current constantly damages motor windings and brushes thermally. Therefore, it is recommended to operate the DC motor with one fourth of 5.6 A. Hence, a 12 V–2 A DC power supply is adequate to feed the DC motor under recommended operation range.

There are three options to satisfy the power settings of the DC motor. The first one is picking a 12 V–2 A adaptor and adjusting its cables to be safely used in operation. The second option is using an available power source such as https://www.sparkfun.com/products/15701. The third option is using an adjustable power supply such as https://www.digikey.com/product-detail/en/global-specialties/1325/GS1325-ND/7606532. The reader can select one of these three options, or a different one, suitable for his or her needs.

2.3.2 The DC Motor Drive Expansion Board

We need specific circuitry to drive the DC motor introduced in previous section. Therefore, we pick the X-NUCLEO-IHM04A1 DC motor drive expansion board developed by STMicroelectronics. Although there are similar boards in the market, we specifically picked this board since it is compatible with our STM32 board. Besides, it is also compatible with Arduino as well.

The DC motor drive expansion board is as in Figure 2.5. It has a voltage range of 8–50 V (maximum) with DC load current 2.8 A. This means if the motor is driven by a constant DC voltage or PWM signal, maximum load current is 2.8 A DC

Figure 2.5 DC motor drive expansion board. (Source: STMicroelectronics. Used with permission.)

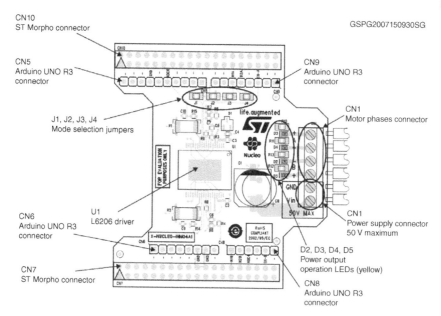

Figure 2.6 General settings of the DC motor drive expansion board. (Source: (STM 2015). Used with permission of STMicroelectronics.)

or rms, respectively. More information on the board can be found in http://www .st.com/en/ecosystems/x-nucleo-ihm04a1.html. In order to use this motor drive expansion board with our STM32 board, the second upper pin of the connector CN9 must be connected to the fourth upper pin of the connector CN9 after driver board is placed on the STM32 board.

The DC motor drive expansion board can be used in different settings, such as driving up to two bidirectional motors, up to four unidirectional motors, one high-power bidirectional motor or two high power unidirectional motors. This is done by appropriate connections and making jumper settings which can be found in STM (2015). We provide general layout of the board in Figure 2.6.

As our DC motor will operate under one fourth of 5.6 A, we will use the connection diagram in Figure 2.7 throughout the book. Here, jumpers J1–J4 on the DC motor drive expansion board must be disconnected.

2.3.3 Encoder

The selected DC motor has an integrated encoder as mentioned in the previous section. This is used to measure speed of the motor. The encoder we are using is of type quadrature encoder with 48 counts per revolution (cpr).

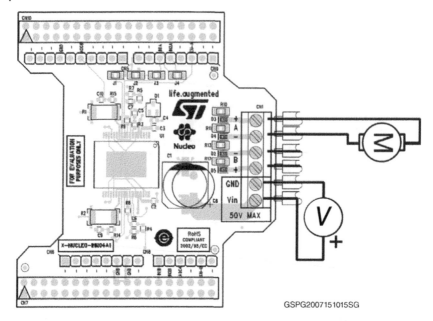

Figure 2.7 Connection between the DC motor drive expansion board and DC motor. (Source: (STM, 2015). Used with permission of STMicroelectronics.)

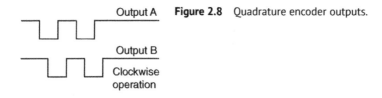

Output A **Figure 2.8** Quadrature encoder outputs.

Output B

Clockwise operation

The quadrature encoder is based on the magnetic two-channel hall effect sensor. This sensor detects the magnetic field of a magnetic disk placed on the motor shaft as it rotates. This creates two square waves as in Figure 2.8.

The hall effect sensor requires an input voltage, V_{CC}, between 3.5 and 20 V and it draws a maximum 10 mA current. The A and B outputs are square waves from 0 V to V_{CC} and they are approximately 90° out of phase. Speed of the motor can be calculated using the frequency of output signals. Direction of the motor can also be obtained from the order of these signals. If both the rising and falling edges of output signals are counted, the sensor provides the resolution of 48 cpr. If the single edge of one output signal is counted, the sensor provides the resolution of 12 cpr. To find the resolution at the output of the gearbox shaft, the sensor cpr must be multiplied by gear ratio which is $48 \times 74.83 = 3591.84$ cpr.

Figure 2.9 FT232 UART to USB converter module. (Source: https://www.hobbypcb.com/index.php/products/accessories/ftdi232.)

2.3.4 The FT232 Module

We will be using the FT232 UART to USB module to send data from the STM32 microcontroller to PC. This module is shown in Figure 2.9. To use it, the reader should connect the PD5 pin of microcontroller to RX pin of FT232, PD6 pin of the microcontroller to TX pin of FT232, and any ground pin of the microcontroller to ground pin of FT232.

As the connection between the microcontroller and the FT232 module is established, the user should open "Device Manager → Ports (COM & LPT) on the PC side. Then, he or she should right click on related COM port shown as "STMicroelectronics STLink Virtual COM Port" and select "Properties." In the opening window, the reader should open "Port Settings" and change "Bits per second" to 921600 as in Figure 2.10. Now, the FT232 module is ready for operation.

2.4 Systems and Sensors to be Used in Advanced Applications

The aim of this book is applying digital control methods to solve real-life problems. In solving these problems, we can benefit from actual systems and sensors which can be purchased from an electronic supplier website. Here, we summarize possible systems and sensors. Hence, the reader will have a general knowledge on them.

2.4.1 Systems

There are wide variety of modules which can be used as a system. We tabulate the most suitable ones for digital control applications in Table 2.2. In the same

Figure 2.10 Baud rate setting of the FT232 module. (Source: Used with permission from Microsoft.)

Table 2.2 Systems to be used in advanced applications.

System name	Sample usage area
Buzzer	Sound generation
DC motor without encoder	Line following robot
Heater table	3D printer
Linear actuator	Generating linear motion
Peltier	Cooling nearby objects
Servo motor	Position control
Solenoid	On–off control
Step motor	Robot position control
Three-phase motor	Drones
Vibration motor	Generating vibration
Voice coil	Vibration cancellation
Water pump	Irrigation

Table 2.3 Sensors to be used in advanced applications.

Sensor name	Sample usage area	Output
Air pressure sensor	Measuring barometric pressure	Digital
Analog accelerometer	Measuring acceleration	Analog
Digital accelerometer	Measuring acceleration	Digital
Distance measuring sensor	Measuring distance	Analog
Encoder	Measuring speed of a motor	Digital
External temperature sensor	Measuring temperature	Digital
Force sensitive resistor (FSR)	Measuring applied force	Analog
Gyroscope	Measuring slope	Analog
Joystick	Measuring position in xy axis	Analog
Light dependent resistor (LDR)	Measuring light intensity	Analog
Load cell	Measuring weight	Analog
NIR LED pair	Line following robot	Analog
Soil moisture sensor	Measuring soil moisture level	Analog
Vibration sensor	Measuring vibration	Analog
Water level sensor	Measuring water level	Analog
Water flow sensor	Measuring water flow	Digital

table, we also provide a brief explanation of the sample usage area of each system. While picking these, we paid attention to their general availability. Please use these systems with care since some of them (such as heater or drone motor with propeller) may be harmful when used without caution. Besides, usage of these possible dangerous systems are not mandatory to understand the concepts explained throughout the book.

2.4.2 Sensors

We also have candidate sensors to be used in control applications. We tabulate them in Table 2.3. These sensors are also picked such that they can be purchased easily. To note here, usage area of each sensor is not limited to the one tabulated in Table 2.3. They can be used in other applications as well.

2.5 Summary

This book aims to blend theory with practice for digital control concepts. Therefore, we focused on the hardware to be used throughout the book in this chapter.

To do so, we started with properties of the STM32 microcontroller and STM32 board which will be embedded hardware environments for implementation. Then, we explored the DC motor and its driver board as basic systems to be used. Related to these, we introduced systems and sensors which can be used in advanced applications. We will be extensively using the hardware introduced in this chapter. Therefore, the reader can consult this chapter whenever needed.

Problems

2.1 Why do we need a development board instead of the microcontroller alone?

2.2 Provide pin names of the STM32 microcontroller.

2.3 Can a pin of the STM32 microcontroller be used for more than one purpose? If this is the case, give one such example.

2.4 What is the voltage range the STM32 microcontroller can operate?

2.5 How are the supply and ground voltages named within the STM32 microcontroller?

2.6 What is the method to program the STM32 microcontroller?

2.7 Which methods can be used to power the STM32 board?

2.8 Which architecture is the STM32 microcontroller based on?

2.9 Why do we need a clock signal for the CPU and peripheral units within the microcontroller?

2.10 What is the difference between the flash and RAM? Provide the size of these two modules within the STM32 microcontroller.

2.11 How many timer modules does the STM32 microcontroller has? What is the reason for having different timer modules within the microcontroller?

2.12 What does ADC and DAC stand for?

2.13 What is the difference between the DAC and PWM signals?

2.14 Which digital communication modules are available within the STM32 microcontroller?

2.15 Summarize physical characteristics of the DC motor to be used throughout the book.

2.16 Why do we need a DC motor expansion board for?

2.17 Explain working principles of the encoder on the DC motor.

2.18 Why will we need the FT232 module throughout the book?

2.19 Give names of at least three systems those can be used in advanced applications. Summarize their sample usage.

2.20 Give names of at least three sensors those can be used in advanced applications. Summarize their sample usage.

3

Software to be Used in the Book

We will approach digital control algorithm implementation in three levels as high, middle, and low. Python running on PC will be the medium for high-level implementation. It will allow the reader to implement and test the desired control algorithm offline. Besides, it will help the reader to visualize the system and signal of interest. Hence, they can be analyzed easily. MicroPython (Python for microcontrollers) will be the medium to be used in middle-level implementation. We can think of MicroPython as the intermediate step between PC and STM32 microcontroller implementations. Here, we will benefit from properties of the Python programming language on the microcontroller. However, this implementation will not be optimal in terms of microcontroller resource usage. Therefore, low-level implementation of digital control algorithms will be done by C language on the STM32 microcontroller. To do so, we will benefit from Arm Mbed Studio to program and debug the C code on the microcontroller. Neither MicroPython nor Mbed Studio is specific to the STM32 microcontroller. Hence, the reader can migrate the codes given in this book to another hardware platform that support MicroPython or Mbed Studio. Therefore, the book content has a wide range of implementation options.

In the following sections, we will cover Python, MicroPython, and C-based implementation steps in detail. We expect the reader to master them. Hence, methods introduced in this chapter can be used in the following chapters for digital control algorithm implementation. As the end-of-chapter application, we will introduce how a DC motor can be run by the STM32 microcontroller via Python and C codes. This will be the backbone of applications to be covered in the following chapters.

Embedded Digital Control with Microcontrollers: Implementation with C and Python,
First Edition. Cem Ünsalan, Duygun E. Barkana, and H. Deniz Gürhan.
© 2021 The Institute of Electrical and Electronics Engineers, Inc. Published 2021 by John Wiley & Sons, Inc.
Companion website: www.wiley.com/go/Unsalan/Embedded_Digital_Control_with_Microcontrollers

3.1 Python on PC

Python is a prototyping language used in scientific and engineering applications. We will be using Python 3 on PC throughout the book. We leave the Python installation and downloading steps to the reader. To note here, a text editor will be needed to modify Python files on PC.

Python has several add-on libraries provided by the user community free of charge. Although these libraries simplify life for the user, we will avoid them whenever possible since they are not available in MicroPython. Hence, we will only use the mandatory Python libraries on the PC side. For all other operations, we will use the basic Python commands which are available for both PC and microcontroller implementations.

There are excellent books on Python programming language. Here, our aim is not to cover Python as they did. We will only focus on fundamental topics of interest for our book. Therefore, let us introduce them next.

3.1.1 Basic Operations

The first stage in Python is checking whether the environment is working or not. We can test this by simply printing a string on the Python command window. To do so, we should use the `print` function as in Listing 3.1.

Listing 3.1: Basic operations in Python.

```
# Test the environment

print("Hello world!")

# Basic arithmetic operations

a=4+3
print(a)

b=1.2-2.3
print(b)

c=4.5*3
print(c)

d=4/3
print(d)

e=4**3
print(e)
```

We can apply basic arithmetic operations on both integer and float numbers in Python. We provide samples on this topic in Listing 3.1. Here, we applied addition, subtraction, multiplication, division, and power operations, respectively. As can be seen in this code, Python has a simple syntax for variables and arithmetic operations.

3.1.2 Array and Matrix Operations

The second stage in Python is getting familiar with array and matrix (more specifically list) usage. These topics are extremely important for us since we will represent a digital signal as an array in Python. Likewise, we will benefit from matrices in representing state-space forms. We provide samples on this topic in Listing 3.2. Here, we start with basic array operations. We also added the visualization option for arrays in Listing 3.2. To do so, we imported the `matplotlib` library in our code. We will use this library throughout the book to plot arrays of interest. As can be seen in the code, there are two plot options. First, we can use the `plot` command to have continuous result. Second, we can use the `stem` command to emphasize digital nature of the signal. We can benefit from the `figure` function to plot arrays in different figures. Finally, we can use the `show` function to display all figures.

Listing 3.2: Array and matrix operations in Python.

```python
#Array operations
A=[2,4,6,8]
print(A[0])
print(A[2])
print(A)
print(A[1:3])

#Duplicate the array
B=A.copy()

B.append(10)

C=A+B
print(C)

#Copy the array
B=A
print(A)

B.append(10)
print(A)

#Define a list of numbers as array
n=list(range(0, 5))
print(n)

#Apply algebraic operations on the array
C=list(map(lambda x:x**2,A))
print(C)

#Plotting array elements
import matplotlib.pyplot as plt

plt.figure(1)
plt.plot(B,'o')
plt.ylabel('value')
plt.xlabel('index')
```

```
plt.figure(2)
plt.stem(n,B)
plt.show()

#Matrix operations
N = 3
A = [[None] * N for i in range(N)]

print(A)

for row in range(N):
    for col in range(N):
        A[row][col] = row + col

print(A)

A[1][1]=12

print(A)
```

The numpy library provides extensive options for array and matrix usage in Python. Unfortunately, we cannot include this library in MicroPython. Hence, we should define the matrix by loop operations (to be explored next). Moreover, we cannot directly apply algebraic operations such as addition and multiplication on array and matrix representations (when the numpy library is not used). Therefore, the reader should form the structure to make algebraic operations on arrays and matrices.

3.1.3 Loop Operations

The third stage in Python is getting familiar with loop operations. This topic is important since we will process arrays (hence signals) in digital control applications via loop operations most of the times. We provide samples on this topic in Listing 3.3. Here, we provide examples on loops constructed by for and while keywords. As can be seen in this code, Python has a unique syntax for loop operations such that operations within the loop are indicated by a right tab. This will be the case for conditional statements and function definitions as well.

Listing 3.3: Loop operations in Python.

```
import matplotlib.pyplot as plt

# for usage
A=[8,4,6,2]
B=[]

for a in A:
    print(a)
```

```
        square=a**2
        B.append(square)

plt.stem(B)
plt.ylabel('value')
plt.xlabel('index')
plt.title('Squared')
plt.show()


# while usage
A=[8,4,6,2]
n=0

while n<len(A):
        print(A[n])
        n+=1
```

3.1.4 Conditional Statements

The fourth stage in Python is getting familiar with conditional statements. We provide samples on this topic in Listing 3.4. Here, we benefit from the if keyword. As can be seen in this code, we again use the right tab to indicate operations within conditional statements.

Listing 3.4: Conditional statements in Python.

```
import matplotlib.pyplot as plt

A=[-1,2,3,-1,2,-4]
B=A[:]

# if else usage
for n in range(0,len(A)):
        print([n,A[n]])
        if A[n]>0:
                B[n]=A[n]
        else:
                B[n]=abs(A[n])

plt.stem(B)
plt.title('Selection')
plt.show()
```

3.1.5 Function Definition and Usage

The fifth stage in Python is getting familiar with function definition and usage. We provide samples on this topic in Listing 3.5. Here, we form a simple function to operate on an array. As in loop operations and conditional statements, commands within the function are indicated by right tab.

Listing 3.5: Function definition and usage in Python.

```
#function definition and usage
def funct_abs(A):
    B=A[:]
    for n in range(0,len(A)):
        B[n]=abs(A[n])
    return B

A=[-1,2,3,-1,2,-4]

C=funct_abs(A)
print(A)
print(C)
```

3.1.6 File Operations

The sixth and final stage in Python is getting familiar with file usage. We provide samples on this topic in Listing 3.6. Here, we read data from a file, process it, and write it back to a new file. To note here, file operations are restricted to the PC environment only.

Listing 3.6: File operations on PC.

```
#File operations, specific to PC

#reads as string, not integer
with open('values.txt') as file_values:
        C=file_values.read()
print(C)

#reads as string, convert to integer, apply operation
D=[]
with open('values.txt') as file_values:
        C=file_values.readlines()

for c in C:
        D.append(2*int(c.rstrip()))

print(D)
print(D[2])

#writes the result to new file
filename='values_modified.txt'
with open(filename,'w') as file_object:
        file_object.write(str(D))
```

3.1.7 Python Control Systems Library

Although we try to avoid add-on libraries in Python, there is one which we cannot escape using on PC. This is the Python control systems library offered in the website https://python-control.readthedocs.io/en/0.8.2/. This is the most comprehensive and complete library on control functions as we were preparing this book.

Being open source, it is being updated and expanded regularly. Therefore, we benefit from it on the PC side whenever needed throughout the book.

The reader can install the Python control systems library by the `pip install control` command. Please see the library website for more recent information on the installation process. We can import this library by adding the `import control` command at the beginning of our Python code on PC. We will be providing usage examples of the Python control systems library in the following chapters.

3.2 MicroPython on the STM32 Microcontroller

MicroPython is an open source programming language based on Python 3 that is optimized to run on selected microcontrollers. The reader can write a Python code and execute it on the microcontroller with MicroPython. Therefore, we will benefit from MicroPython as the intermediate step between the Python-based PC implementation and C-based STM32 microcontroller implementation. Next, we explore it in detail.

3.2.1 Setting up MicroPython

The latest MicroPython version for the STM32 microcontroller can be downloaded from https://micropython.org/download. The reader should check the "STM32 boards" section and select the appropriate board file with the `.dfu` extension in this website. Due to the hardware used in this book, we had to recompile our own MicroPython version. We provide it in the accompanying website. However, we provide the standard installation steps here to be compatible with the future MicroPython versions.

There are two ways to load the `dfu` file to the microcontroller. The first method is downloading the STSW-STM32080 pack from http://www.st.com. In this pack, the reader will find the DfuSeDemo program to load `dfu` files to STM microcontrollers. In order to use it, we need to start microcontroller in DFU mode by connecting BOOT0 pin (7th pin of CN1 connector) to VDD pin (5th pin of CN1 connector) using a jumper. After a reset, the reader will see "STM Device in DFU Mode" notification in "Available DFU Devices" window. General layout for this window is as in Figure 3.1. Afterward, the reader should click "Choose" and select the related `dfu` file. Finally, the reader should click "Upgrade" and wait until the loading process is finished.

The second method in loading the `dfu` file to the STM32 microcontroller is downloading STSW-STM32080 and STSW-LINK004 packs from http://www.st .com. In the STSW-LINK004 pack, the reader will find "STM32 ST-LINK Utility" program to load `hex` files to the STM32 microcontroller. In the STSW-STM32080

Figure 3.1 Available dfu devices window. (Source: Used with permission from Microsoft.)

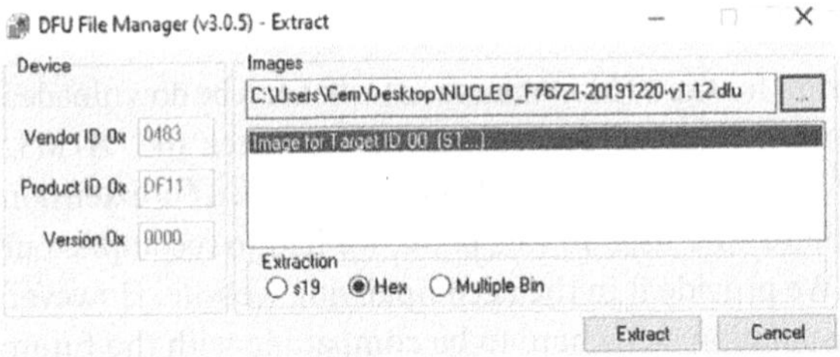

Figure 3.2 Selecting the related dfu file. (Source: Used with permission from Microsoft.)

pack, the reader will find "DFU file manager" program to convert dfu files to hex files. To do so, the reader should first open DFU file manager program and select "I want to EXTRACT S19, HEX or BIN from a DFU one." In the opening window, he or she should select the related dfu file and select "Hex Extraction" as in Figure 3.2.

When the reader clicks "Extract" in Figure 3.2, a hex file will be created in the same directory. Then, the reader should open the STM32 ST-LINK Utility program and select "Target, Program & Verify." In the opening window, he or she should browse the related hex file as in Figure 3.3. Here, the reader should use the provided hex file in the book website. Finally, the reader should click "Start" and wait until the loading process is finished.

Before explaining how to program the STM32 microcontroller using Micro-Python, there is a good online simulator at https://micropython.org/unicorn/. The simulator uses the PyBoard instead of the STM32 board, but it can be

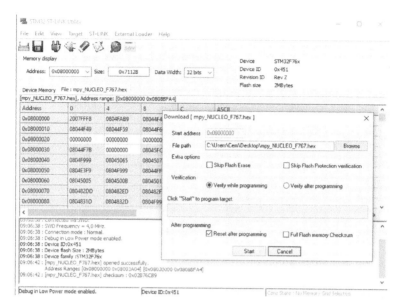

Figure 3.3 Browsing the related .hex file. (Source: Used with permission from Microsoft.)

useful to learn basic concepts of MicroPython. Also, detailed documentation on MicroPython can be found at http://docs.micropython.org/en/latest/.

3.2.2 Running MicroPython

After setting up MicroPython, there are two methods to use it. The first method is based on MicroPython's unique property called read evaluate print loop (REPL). The reader can access the MicroPython REPL over USB serial connection. This allows connecting to the board and executing the code without any need for compiling or uploading it to the microcontroller. This is perfect for experimenting with hardware. In order to access REPL, the reader should open a terminal program (such as Tera Term) on PC and connect to the serial COM port associated with the STM32 board. The port name can be found in "Device Manager" under Windows, with the name "STMicroelectronics STLink Virtual COM Port." Here, the reader should set the baud rate to 115200 bps. The opening window will be as in Figure 3.4.

The reader may need to restart MicroPython by pressing the reset button on the board. The `help()` command can be used to list primary help topics. Also, the tab button can be used to see the content of a module. For example, when the reader types `pyb.` and presses the tab button, he or she can see all objects related to the `pyb` module.

Figure 3.4 Tera Term opening window. (Source: Used with permission from Microsoft.)

The second method in using MicroPython is by accessing the `main.py` file. MicroPython is directly loaded to the flash of the microcontroller. Here, there are two files as `boot.py` and `main.py` to control the execution of Python code. The `boot.py` file is called by the system and it contains initial setup commands. After setup, it automatically calls the `main.py` file. Hence, we need to change the content of the `main.py` file to execute our program.

There are two ways to change content of the `main.py` file. The first method is using a micro USB connection. When the reader connects the micro USB cable to the STM32 board and loads MicroPython as described previously, he or she can directly access flash memory of the microcontroller and change the `boot.py` and `main.py` files using a text editor. However, this method may cause a corrupt file problem.

The second method in changing the content of the `main.py` file is using the special MicroPython shell called `rshell`. Detailed information for this shell can be found in https://github.com/dhylands/rshell. In order to use `rshell`, the user should install it by using the command `pip3 install rshell`. Here, we suggest installing rshell to the scripts subfolder of the running Python folder on PC. After installation, it can be started using the `rshell` command. Then, the user can connect to the STM32 board using the command `connect serial COMx` on the command prompt. Here x stands for the COM port number the STM32 board is connected to. As connection to the STM32 board is established, content of the `main.py` file on the microcontroller can be modified by using the command `cp Source /flash/main.py`. Here, the Python code written in the `Source` file is copied to the `main.py` file. The reader should also add path of the file in operation. Also, other MicroPython files can be copied

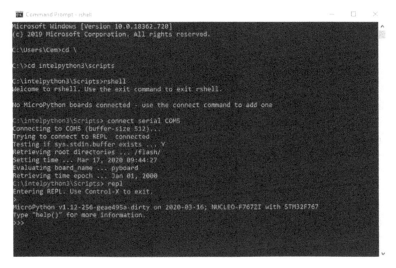

Figure 3.5 REPL in the rshell command window. (Source: Used with permission from Microsoft.)

to microcontroller flash using `cp Source /flash` command. Again, path of the file must be provided here. After the desired files are copied, the STM32 microcontroller must be reset by the button on the STM32 board. The user can also enter the MicroPython REPL over `rshell` using the command `repl`. We provide the sample command window content as a result of this operation in Figure 3.5. The reader can exit from REPL using the command `Ctrl-x`. Please see the mentioned website for more information about available REPL shell commands.

When the `main.py` file is corrupted, the reader needs to reset the STM32 microcontroller to its initial settings. To do so, MicroPython's "factory reset the file system property" can be used. Internal flash of the microcontroller is returned to its initial state in this mode. To start MicroPython in this mode, press both the user and reset buttons. Then, release the reset button. LEDs are turned on in a cycle as only blue LED, only red LED, and both blue and red LEDs. When both LEDs are turned on, release the user button. The blue and red LEDs should flash quickly four times. Then the green, red, and blue LEDs should be turned on. Wait until all LEDs are turned off and MicroPython starts in factory reset mode.

We can test whether all the above steps have been implemented correctly by adding the `print(''Hello world!'')` command in the `main.py` file. As the code is executed, we expect to see the string `Hello world!` in the REPL window. The reader should check all the above steps when the string cannot be observed. From this point on, we will assume that MicroPython works without any problem on the STM32 microcontroller.

3.2.3 Reaching Microcontroller Hardware

Since MicroPython runs on the microcontroller, it is important to reach micro-controller hardware using it. Fortunately, this can be achieved by adding specific modules to our Python code. We will evaluate these next.

3.2.3.1 Input and Output Ports

STM32 microcontroller general purpose input and output (GPIO) pins can be accessed by using the `pyb` module, and they will be represented by an object named `Pin`. Here, we describe the GPIO pin operations using the `pyb.Pin` function. We need to initialize the GPIO pin first with `pyb.Pin('PinName', mode=PinMode, pull=PinPushPull, af=PinAlternateFunction)`. Here, the `PinName` can be entered as `Pxy` with x being the port name and y being the pin number. `PinMode` is used to set pin as input or output and can be one of `Pin.IN`, `Pin.OUT_PP` (Push Pull), `Pin.OUT_OD` (Open Drain), `Pin.ANALOG`, `Pin.AF_PP` (Alternate Function Push Pull) or `Pin.AF_OD` (Alternate Function Open Drain) depending on the usage type. `PinPushPull` is used to enable or disable pull up/down resistors and can be `Pin.PULL_NONE`, `Pin.PULL_UP` or `Pin.PULL_DOWN`. `PinAlternateFunction` is used to select an alternate function for the related pin. These alternate functions can be observed using the function `Pin.af_list()`. For example, if the user wants to see the alternate functions of PA0 pin, he or she should use the command `pyb.Pin(PA0).af_list()` and writes the desired alternate function name to `PinAlternateFunction`.

If a pin is initialized as output, its value can be set using the command `Pin.value(result)`. Here, `result` can be logic level 1 or 0. The pin can also be directly set to logic level 1 using the command `Pin.high()` or `Pin.on()`. It can also be set to logic level 0 using the function `Pin.low()` or `Pin.off()`. A sample code using these functions can be found in Listing 3.7. Here, the blue LED connected to pin PB7 on the STM32 board toggles every second. We used the `pyb.delay(ms)` function in this sample code to generate the desired delay.

Listing 3.7: Toggle the onboard blue LED in MicroPython.

```python
import pyb

def main():
        LD2 = pyb.Pin('PB7', mode=pyb.Pin.OUT_PP)

        while True:
                LD2.value(1)
                pyb.delay(1000)
                LD2.value(0)
                pyb.delay(1000)

main()
```

If a pin is initialized as input, its value can be observed using the command Pin.value(). A sample code using this function can be found in Listing 3.8. Here, the onboard green LED connected to pin PB0 is turned on when the onboard push button connected to pin PC13 is pressed and turned off when the button is released.

Listing 3.8: Turn on and off the onboard green LED using the onboard push button in MicroPython.

```
import pyb

def main():
        LD1 = pyb.Pin('PB0', mode=pyb.Pin.OUT_PP)
        B1 = pyb.Pin('PC13', mode=pyb.Pin.IN, pull=pyb.Pin.PULL_NONE)

        while True:
                LD1.value(B1.value())

main()
```

3.2.3.2 Timers

The timer of the STM32 microcontroller can be accessed from the pyb module using the object Timer. To use the timer, first it should be initialized using the function pyb.Timer(TimerID, freq=TimerFrequency, mode=TimerMode). Here, TimerID is the number of the desired timer. TimerFrequency is the desired frequency value, and all timer registers are arranged automatically to create this frequency. The TimerMode can be selected as Timer.UP, Timer.DOWN, or Timer.CENTER. These define the counting order whether it will be from zero to AutoReload Register (ARR) value, or from ARR to minimum value or counting up then down. The ARR value is calculated automatically according to the selected timer frequency. Also, the Timer.callback() function can be used to define what will happen when the timer counts to the desired value. A sample code using timer functions can be found in Listing 3.9. Here, the onboard red LED connected to pin PB14 is toggled every two seconds using the Timer2 interrupt.

Listing 3.9: Toggling the onboard red LED using the Timer2 interrupt in MicroPython.

```
import pyb

LD3 = pyb.Pin('PB14', mode=pyb.Pin.OUT_PP)

def toggle_LD3(timer):
        LD3.value(not LD3.value())

def main():
        Timer2 = pyb.Timer(2, freq=0.5, mode=pyb.Timer.UP)
        Timer2.callback(toggle_LD3)

main()
```

External pins can also be configured to generate pulse width modulation (PWM) signals. The function pyb.Timer.channel(TimerChannel, mode=ChannelMode, pin=pyb.Pin('PinName')) can be used for this purpose. Here, TimerChannel is the channel number of the desired timer. ChannelMode should be selected as Timer.PWM. Although there are other options here, we will just focus on PWM usage. The PinName can be entered as Pxy with x being the port name and y being the pin number. When the mode is selected as PWM, there is an additional input called pulse_width_percent to determine the duty cycle.

Sample code using timer functions for PWM generation can be found in Listing 3.10. Here, brightness of the green LED connected to pin PB0 is controlled using PWM. When the onboard push button connected to pin PC13 is pressed, duty cycle value is increased by 10, between 0 and 100, inside the external interrupt callback function. The function pyb.ExtInt('PinName', mode=TriggerMode, pull=PinPushPull, callback=PinCallback) is used to create external interrupt. Here, the PinName can be entered as Pxy with x being the port name and y being the pin number. TriggerMode is used to select trigger edge and can be one of ExtInt.IRQ_RISING, ExtInt.IRQ_FALLING or ExtInt.IRQ_RISING_FALLING. PinPushPull is used to enable or disable pull up/down resistors and can be Pin.PULL_NONE, Pin.PULL_UP or Pin.PULL_DOWN. PinCallback is the name of created external interrupt callback function. Finally, initial duty cycle is set to 0 and frequency of the PWM signal is set to 1 kHz.

Listing 3.10: Controlling brightness of the onboard green LED using PWM in MicroPython.

```
import pyb

Timer3 = pyb.Timer(3, freq=1000, mode=pyb.Timer.UP)
Timer3_PWM = Timer3.channel(3, mode=pyb.Timer.PWM, pin=pyb.Pin('PB0'),
    pulse_width_percent=0)
duty_cycle = 0

def increase_duty_cycle(pin):
        global duty_cycle
        duty_cycle = duty_cycle + 10
        if duty_cycle > 100:
                duty_cycle = 0
        print(duty_cycle)
        Timer3_PWM.pulse_width_percent(duty_cycle)

def main():
        pyb.ExtInt('PC13', mode=pyb.ExtInt.IRQ_RISING, pull=pyb.Pin.
            PULL_NONE, callback=increase_duty_cycle)

main()
```

3.2.3.3 ADC

We will extensively use analog to digital conversion (ADC) operation throughout the book. We will also introduce theoretical aspects of this operation in Chapter 5. We can use ADC module of the STM32 microcontroller via MicroPython. To do so, we should initialize the ADC module using the function `pyb.ADC(pin=pyb.Pin('PinName'))`. Here, `pin` is the microcontroller pin with analog input capability. The `PinName` can be entered as `Pxy` with `x` being the port name and `y` being the pin number. Then, analog value from this pin can be read using the function `ADC.read()` in 12-bit resolution. Whenever this function is called in the code, it performs single read. ADC input voltage can be calculated using

$$Voltage = \frac{3.3 \times ADC\_value}{4095} \qquad (3.1)$$

A sample code using ADC functions can be found in Listing 3.11. Here, repeated reads from the pin PA3 are converted to voltage and printed in a while loop. Sampling frequency for this process is set to 2 Hz using the function `pyb.delay()`. Also, the onboard green LED connected to pin PB0 turns on if the ADC voltage is higher than 1.5 V and turns off otherwise.

Listing 3.11: Getting analog value from pin PA3 using polling in MicroPython.

```python
import pyb

def convert_ADCvalue_to_voltage(value):
        voltage = 3.3 * value / 4095
        return voltage

def main():
        adc_A3 = pyb.ADC(pyb.Pin('PA3'))
        LD1 = pyb.Pin('PB0', mode=pyb.Pin.OUT_PP)

        while True:
                ADC_value = adc_A3.read()
                ADC_voltage = convert_ADCvalue_to_voltage(ADC_value)
                print("ADC voltage is %6.4f"% (ADC_voltage))
                if ADC_voltage > 1.5:
                        LD1.value(1)
                else:
                        LD1.value(0)
                pyb.delay(500)

main()
```

The ADC module also has the function `ADC.read_timed(buf, timer)` to perform multiple reads with a precise sampling rate. Here, number of ADC reads is performed until the `buf` array is full. The specified sampling frequency is created by the timer module. However, this is a blocking function. Therefore, code execution is halted until the `buf` is full. Therefore, the timer callback function can be used instead to perform a nonblocking operation. A sample code for this operation is given in Listing 3.12. Here, consecutive 20 reads from pin PA6 are saved to the buffer array in the timer callback function. The sampling frequency for this process is set to 10 Hz using Timer1. The timer interrupt is disabled after 20 reads. Then, stored ADC values are converted to voltage and printed in the main function. Also the onboard blue LED, connected to pin PB7, turns on if the average of read voltage values is larger than 1.5 V and turns off otherwise.

Listing 3.12: Getting analog values from pin PA6 using timer interrupt in MicroPython.

```python
import pyb

from array import array
Timer1 = pyb.Timer(1, freq = 10, mode = pyb.Timer.UP)
adc_A6 = pyb.ADC(pyb.Pin('PA6'))
interrupt_cnt = 0
print_flag = 0
buffer = array('I', (0 for i in range(20)))

def convert_ADCvalue_to_voltage(value):
        voltage = 3.3 * value / 4095
        return voltage

def get_ADC(timer):
        global interrupt_cnt
        global print_flag
        global buffer
        buffer[interrupt_cnt] = adc_A6.read()
        interrupt_cnt = interrupt_cnt + 1
        if interrupt_cnt == 20:
                Timer1.deinit()
                print_flag = 1

def main():
        global print_flag
        sum = 0
        LD2 = pyb.Pin('PB7', mode=pyb.Pin.OUT_PP)
        Timer1.callback(get_ADC)

        while True:
                if print_flag == 1:
                        for i in range(len(buffer)):
                                ADC_voltage =
                                        convert_ADCvalue_to_voltage
                                        (buffer[i])
                                sum = sum + ADC_voltage
                                print(ADC_voltage)
                        print('Printing is done !')
                        print_flag = 0
                        average = sum / 20
                        if average > 1.5:
                                LD2.value(1)
                        else:
                                LD2.value(0)
main()
```

The ADC module can also be used to measure internal temperature sensor, reference voltage, or battery voltage. To do so, we should initialize the related ADC channel using the function `pyb.ADCAll(resolution, mask)`. Here, `resolution` is the ADC resolution and `mask` is the value which is used to select the desired ADC channel. Afterward, `read_core_temp()`, `read_core_vbat()`, `read_vref()`, and `read_core_vref()` functions can be used to read the desired value. A sample code for reading internal microcontroller temperature this way is given in Listing 3.13. Here, internal temperature value is obtained in Celsius degree and printed in an infinite loop every 0.1 second.

Listing 3.13: Getting internal microcontroller temperature in MicroPython.

```
import pyb

def main():
        adc_temp = pyb.ADCAll(12, 0x70000)
        while True:
                value = adc_temp.read_core_temp()
                print(value)
                pyb.delay(100)

main()
```

3.2.3.4 DAC

We will also extensively use digital to analog conversion (DAC) operation as with ADC throughout the book. We will introduce theoretical aspects of this operation in Chapter 5. We can use DAC module of the STM32 microcontroller via MicroPython. To do so, we should initialize the DAC module using the function `pyb.DAC(port, bits=DACbits, buffering=DACbuffer)`. Here, `port` is the predefined DAC output pin and can be selected as 1 or 2. Based on the selection, pin PA4 or PA5 can be used for DAC output, respectively. `DACbits` is used for selecting resolution of the DAC module. This value can be 8 or 12 for the related resolution. `DACbuffer` is used to enable or disable the output buffer. It can be enabled using `True` or disabled using `False`. After the DAC module is initialized, the function `DAC.write(value)` can be used to feed the desired value to DAC output. Here, `value` can be selected between 0 and $2^{bits} - 1$. The DAC output voltage can be calculated using

$$Voltage = \frac{3.3 \times value}{4095} \tag{3.2}$$

A sample code using DAC functions can be found in Listing 3.14. Here, the DAC output value increased by 0.01 V every 10 ms until it reaches 3.3 V. Then, the DAC output is reset to 0 V again. The output can be observed by connecting a multimeter to pin PA4.

Listing 3.14: DAC output from pin PA4 using polling in MicroPython.

```python
import pyb

def convert_DACvoltage_to_value(voltage):
        value = int(4095 * voltage / 3.3)
        return value

def main():
        DAC_voltage = 0
        dac1 = pyb.DAC(1, bits=12, buffering=True)

        while True:
                result = convert_DACvoltage_to_value(DAC_voltage)
                dac1.write(result)
                DAC_voltage = DAC_voltage + 0.01
                if DAC_voltage > 3.3:
                        DAC_voltage = 0
                pyb.delay(10)

main()
```

There is yet another function for controlling the DAC output. This is DAC.write_timed(data, freq=DACfrequency, mode=DACmode) which uses the direct memory access (DMA) module to feed the values inside an array to DAC output. Here, data is the array that holds consecutive DAC values. Type of this array must be unsigned short ('H'). DACfrequency is the frequency of the DAC module and Timer6 is automatically used to create this frequency. Other timers can also be used to create desired frequency using pyb.Timer function. Finally, DACmode is the DAC output mode. It can be DAC.NORMAL or DAC.CIRCULAR. A sample code using this function is provided in Listing 3.15. Here, voltage values for a sinusoidal signal with 128 elements are created using the array module. Then, they are converted to DAC values and saved in another array. Finally, this array is fed to DAC output continuously in a circular mode. DAC frequency is set to 8 Hz using Timer2. Hence, frequency of the sinusoidal signal is 8/128 Hz. The output can be observed by connecting a multimeter to pin PA4.

Listing 3.15: DAC output from pin PA4 using DMA in MicroPython.

```python
import pyb

from array import array
from math import pi,sin

def convert_DACvoltage_to_value(voltage):
        value = int(4095 * voltage / 3.3)
        return value

def main():
        sine_voltages = array('f', 1.65 + 1.65 * sin(2 * pi * i /
                128) for i in range(128))
        DAC_values = array('H', (0 for i in range(128)))
```

```
for i in range(len(DAC_values)):
        DAC_values[i] = convert_DACvoltage_to_value(
            sine_voltages[i])
dac1 = pyb.DAC(1, bits=12, buffering=True)
dac1.write_timed(DAC_values, pyb.Timer(2, freq=8), mode=dac1
    .CIRCULAR)

main()
```

There are two more DAC functions which can be useful in operations. These are `DAC.noise(freq)` and `DAC.triangle(freq)`. The `DAC.noise(freq)` function is used to generate a pseudo-random noise signal at DAC output for a given frequency. The `DAC.triangle(freq)` function is used to generate a triangle signal at DAC output.

3.2.3.5 UART

Universal asynchronous receiver/transmitter (UART) module of the STM32 microcontroller can be accessed by the `pyb` module using the object `UART`. To use this module, first it should be initialized by the function `pyb.UART(UARTID, BaudRate)`. Here, `UARTID` is the number of the desired UART pin which can take values between 1 and 5. These ID numbers and corresponding UART pins are tabulated in Table 3.1.

The desired text can be sent using the command `UART.write('TEXT')`. Also, the function `UART.writechar()` can be used to send one character. To read data, `UART.read()` function can be used. If input of the function is set null, all available characters are read. If input of the function is an integer value, then number of characters defined by this integer are read. Also, `UART.readchar()` function can be used to read one character.

Sample codes using UART functions can be found in Listings 3.16 and 3.17. The string "Hello world!" is sent to PC from the microcontroller via UART every second in Listing 3.16. The onboard green LED turns on and off in Listing 3.17 by button presses "h" and "l" on the PC keyboard.

Table 3.1 UART pins for ID numbers.

ID	TX pin	RX pin
2	PD5	PD6
3	PD8	PD9
5	PB6	PB12
6	PC6	PC7

Listing 3.16: Sending the string "Hello world!" to PC via UART in MicroPython.

```
import pyb

def main():
        uart3 = pyb.UART(3)

        while True:
                uart3.write('Hello World !\r\n')
                pyb.delay(1000)

main()
```

Listing 3.17: Turning on and off the onboard green LED using the PC keyboard in MicroPython.

```
import pyb

def main():
        uart3 = pyb.UART(3)
        LD1 = pyb.Pin('PB0', mode=pyb.Pin.OUT_PP)

        while True:
                input = uart3.readchar()
                if input == 0x68:
                        LD1.on()
                elif input == 0x6C:
                        LD1.off()

main()
```

3.2.4 MicroPython Control Systems Library

Inspired by the Python control systems library introduced in Section 3.1.7, we constructed a new library to be used under both PC and MicroPython for microcontrollers. We call this as MicroPython control systems library. We should emphasize that Python and MicroPython control systems libraries do not compete. They complete each other. We will also see that there are functions in our library to form a bridge between both libraries. Hence, they can be used together.

Due to memory size constraints on the STM32 microcontroller, we partitioned our MicroPython control systems library into two parts. The first part, called mpcontrolPC.py, can only be used on PC. The second part, called mpcontrol.py, can be used both on PC and STM32 microcontroller. We provided both files in the accompanying book website. These files should be added to the working Python directory on PC. Then, we should import them to our Python code by the commands import mpcontrolPC and import mpcontrol. We should add the file mpcontrol.py to flash folder using rshell to run on the microcontroller. Then, we should import it to the main.py file by using the command import mpcontrol.

3.3 C on the STM32 Microcontroller

C is the most effective language to program a microcontroller in terms of resource usage, speed of execution, and ease of programming. Therefore, it is taken as the defacto programming language for most microcontroller platforms. Here, we assume the reader has sufficient knowledge to cover fundamental C programming techniques. We will only focus on major embedded implementation issues.

In order to program the STM32 microcontroller, we need a platform on the PC side. We picked the Arm Mbed Studio as the integrated development environment (IDE) for this purpose. This IDE allows us forming a project including all necessary support files and user code for the microcontroller.

Mbed is a web-based compiler introduced earlier by Arm. Mbed has several advantages. First, it does not need any installation step to operate. The user only needs to open a free account at the website https://os.mbed.com/ide/. Afterward, the project can be developed on a web-based compiler interface. The generated code is downloaded to PC as a file which can be embedded to hardware by a drag and drop operation. The important point here is that, the hardware should be suitable for this operation. Boards supporting this property are listed in the website https://os.mbed.com/platforms/. These boards are labeled as "Mbed Enabled" and they form the Mbed ecosystem. The second and most important advantage of Mbed is that the code generated for one platform in the Mbed ecosystem can easily be ported to another platform as long as hardware requirements are satisfied. This is possible since the Mbed compiler simplifies and generalizes low-level hardware setup and usage. The third advantage of Mbed is that it allows code sharing between Mbed users. This is done by posting and importing the generated code through the web-based interface. In fact, the user can easily port a code from another user by importing it to his or her workspace. Then, this project can be developed further for a specific purpose.

Mbed Studio is the desktop version of Mbed. The user can benefit almost all properties of Mbed under Mbed Studio. Here, an active internet connection is not required for the compilation process. Besides, the generated code and the project are saved in a local folder on PC. As we were writing this book, Mbed Studio also offered an experimental debug interface to control the generated code. We expect it to be in mature form within a couple of months. Due to its advantages, we pick Mbed Studio instead of Mbed. However, the reader can always use the same code in Mbed as well.

Mbed Studio can be downloaded from the website https://os.mbed.com/studio/. When the download is complete, the reader should follow the steps given there for installation. When the installation is complete, the reader should login to his or her Mbed account. Afterward, the program will launch and the IDE will be as in

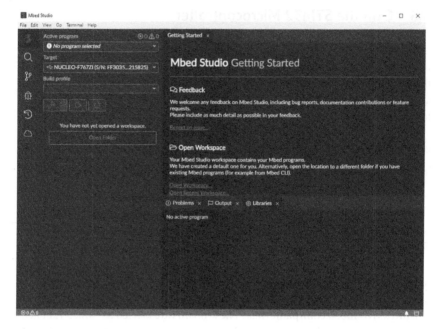

Figure 3.6 Mbed Studio welcome window. (Source: Arm Limited.)

Figure 3.6. We will be using this interface for all our needs. Therefore, let us briefly summarize it.

As can be seen in Figure 3.6, Mbed Studio has a fairly compact interface. On the left side of the window, there exists a panel. Through it, the user can create a new project, select the target, and build the project for the selected target. On the center side of the window, the main code file will be opened as it is added to the project. Hence, the reader will develop his or her code there. On the bottom of the panel, there are tabs which will summarize all build and debug operations. The reader can also select the appearance of the Mbed Studio IDE through the "View" → "Themes" option. From this point on, we select the light theme for the IDE.

3.3.1 Creating a New Project in Mbed Studio

We should form a project in Mbed Studio to program the STM32 microcontroller. A project typically contains source, header, and include files. From these, Mbed Studio generates an executable output file which is used by the microcontroller.

To create a new project, click "File" and "New program." A new window opens up as in Figure 3.7. Here, we must select an "example program" as the template for our project. We will always select the option "mbed-os-example-blinky-baremetal"

Figure 3.7 Forming the new program. (Source: Arm Limited.)

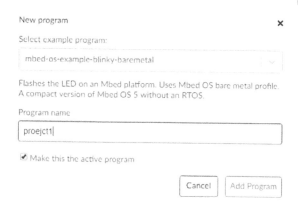

here. Then, we should give a name to our project in the "Program name" window. Afterwards, we should press the "Add Program" button and let Mbed Studio import all necessary libraries.

Mbed Studio can automatically detect the target platform as it is connected to a USB port. As we connect our STM32 board to PC, a new pop-up window should appear asking for "setting device as active?" As we press the "Yes" button, our "NUCLEO-F767ZI" board should be seen in the "Target" selection window on the left panel. The connection sign should also be green to indicate that the board is connected and active.

We can add our C code to the `main.cpp` file under our project. In fact, there is a code in this file since we selected the template project. We should first test whether all settings are done correctly up to now by executing this code. This is done next.

3.3.2   Building and Executing the Code

As we create a new project, the next step is its execution. The first step in executing the code on an embedded platform is building and debugging it. There are two buttons on Mbed Studio's left panel with the hammer and triangle shape to build and run the program, respectively. We can build and execute the program by pressing both. As the program is executed, the green LED on the STM32 board should be blinking every 0.5 seconds. This tells us that the hardware is working properly, and we are ready to go.

3.3.3   Reaching Microcontroller Hardware

We can reach and use all STM32 microcontroller hardware properties through Mbed Studio. Next, we consider input and output ports, timers, ADC and DAC, and digital communication modules step by step as we did in MicroPython.

3.3.3.1 Input and Output Ports

We can modify and use input and output ports via Mbed Studio. At this step, let us assume that we want to turn on and off the green LED at one second intervals. Let us create a simple project for this purpose. To do so, we should set the microcontroller pin settings first. Hence, we should set pin PB0 of the STM32 microcontroller as "GPIO_Output." This modification can be done from Mbed Studio by the code `DigitalOut led1(LED1)`. Here, `LED1` is the predefined constant representing the pin and the green LED on the board is connected to. The variable `led1` will be used throughout the code to reach `LED1`. We can also set pin PB0 as output by the code `DigitalOut led1(PB_0)`. Here, the variable `led1` will indicate the pin `PB_0`. We can use the available delay function `wait_ms(d)` in our program. This function is used for creating a delay for d ms. The sample code generated using these functions is as in Listing 3.18. To execute this code in Mbed Studio, we should copy and paste it to the `main.cpp` file in our active project.

Listing 3.18: Toggle the onboard green LED, the C code.

```
#include "mbed.h"

#define WAIT_TIME 1000 //msec

//DigitalOut led1(LED1);
DigitalOut led1(PB_0);

int main()
{
    while(true){
        led1 = !led1;
        wait_ms(WAIT_TIME);
    }
}
```

In the second example, we want to turn on the green LED whenever the button on the STM32 board is pressed and turn off the green LED when it is released. To do so, we should set pin PC13 of the STM32 microcontroller as "GPIO_Input." As in the previous example, we can perform this operation in two ways. First, we can use the predefined constant `BUTTON1` within the function `DigitalIn button1(BUTTON1)`. Here, we define a variable `button1` to reach `BUTTON1` throughout the code. As in the previous example, we can also directly use the pin address of the button as `DigitalIn button1(PC_13)`. We can set the green LED as output as in the previous example. We provide the C code for this example in Listing 3.19.

Listing 3.19: Turn on and off the onboard green LED using the onboard push button, the C code.

```
#include "mbed.h"
DigitalOut led1(LED1);
DigitalIn button1(BUTTON1);

int main()
{
    while(true){
        led1 = button1;
    }
}
```

3.3.3.2 Timers

We can modify and use timers via Mbed Studio. In the first example, we turn on and off the red LED on the STM32 board at two seconds intervals using the timer interrupt. We provide the corresponding C code in Listing 3.20. Here, we define the timer by `Ticker timer_ticker`. Our timer interrupt function in this example is `toggle_LD3()`. We call this function every two seconds using the code `timer_ticker.attach(&toggle_LD3,2)`.

Listing 3.20: Toggling the onboard red LED using the timer interrupt, the C code.

```
#include "mbed.h"
Ticker timer_ticker;
DigitalOut led3(LED3);

void toggle_LD3(){
        led3=!led3;
}

int main()
{
        timer_ticker.attach(&toggle_LD3, 2);

        while(true);
}
```

In the second example, we want to control brightness of the green LED on the STM32 board using PWM. To do so, we provide the C code in Listing 3.21. Here, we use the definition `PwmOut pwm(LED1)` to feed PWM signal output to the pin green LED is connected to. We set the period and pulse width of the PWM signal by the functions `pwm.period_ms` and `pwm.pulsewidth_us`, respectively. As the code is executed, we can increase brightness of the green LED by increasing the PWM pulse width inside the function `increase_duty_cycle`. This is the callback function for the rising edge interrupt of the user button.

Listing 3.21: Controlling brightness of the onboard green LED using PWM, the C code.

```c
#include "mbed.h"
PwmOut pwm(LED1);
InterruptIn button1(BUTTON1);
int duty_cycle = 0;

void increase_duty_cycle(){
        duty_cycle = duty_cycle + 100;
        if (duty_cycle > 1000)duty_cycle = 0;
        pwm.pulsewidth_us(duty_cycle);
}

int main()
{
        __enable_irq();
    button1.rise(&increase_duty_cycle);
    pwm.period_ms(1);
    pwm.pulsewidth_us(duty_cycle);

        while(true);
}
```

3.3.3.3 ADC

We can modify and use the ADC module via Mbed Studio. In the first example, we acquire analog values from the pin PA3 every 0.5 seconds in polling mode. Then, we display the read value in the terminal available under Mbed Studio. This window can be reached by clicking on the tab with the board name, NUCLEO-F767ZI, at the bottom of the IDE. We provide the C code for this example in Listing 3.22. Here, we define the analog pin as `AnalogIn analog(PA_3)`. We acquire data from this pin using the function `analog.read_u16()`. Then, we convert this value to voltage value using `convert_ADCvalue_to_voltage` function. Finally, we print the acquired voltage value to the terminal window using the function `printf`. In fact, this function sends data to the terminal via UART communication. Also the onboard green LED, connected to pin PB0, is turned on if the voltage value is higher than 1.5 V and turned off otherwise.

Listing 3.22: Getting analog data from pin PA3 using polling, the C code.

```c
#include "mbed.h"

AnalogIn analog(PA_3);
DigitalOut led1(LED1);

float convert_ADCvalue_to_voltage(int value){
        float voltage;
        voltage = 3.3 * ((float)value) / 65535;
        return voltage;
}

int main()
```

```
{
        int ADC_value;
        float ADC_voltage;
    while(true){
    ADC_value = analog.read_u16();
        ADC_voltage = convert_ADCvalue_to_voltage(ADC_value);
    printf("ADC voltage is %6.4f \n", ADC_voltage);
        if(ADC_voltage > 1.5)led1 = 1;
        else led1 = 0;
    wait(0.5);
    }
}
```

In the second example, given in Listing 3.23, consecutive 20 reads from pin PA6 are stored to a buffer array using the ticker callback function. Sampling frequency for this process is set to 10 Hz. The ticker callback function is detached after 20 reads. Then, saved ADC values are converted to voltage and printed in the main function. Also the onboard blue LED, connected to pin PB7, is turned on if average of the read voltage values is higher than 1.5 V and turned off otherwise.

Listing 3.23: Getting analog data from PA6 pin using the timer interrupt, the C code.

```
#include "mbed.h"

Ticker timer_ticker;
AnalogIn analog(PA_6);
int interrupt_cnt = 0;
int print_flag = 0;
int buffer[20];

float convert_ADCvalue_to_voltage(int value){
        float voltage;
        voltage = 3.3 * ((float)value) / 65535;
        return voltage;
}

void get_ADC(){
    print_flag = 1;
}

int main()
{
        int ADC_value;
        float ADC_voltage;
        float average;
        int i;
        float sum = 0;
        DigitalOut led2(LED2);
        timer_ticker.attach(&get_ADC, 0.1);

        while(true){
                if(print_flag == 1){
            if(interrupt_cnt<20){
                    buffer[interrupt_cnt] = analog.read_u16();
                    interrupt_cnt++;
            }
            else{
```

```
timer_ticker.detach();
for(i=0;i<20;i++){
    ADC_voltage = convert_ADCvalue_to_voltage(buffer
        [i]);
    sum = sum + ADC_voltage;
    printf("ADC voltage is %6.4f \n",ADC_voltage);
}
printf("Printing is done ! \n");
average = sum / 20;
if(average > 1.5)led2 = 1;
else led2 = 0;
}
        print_flag = 0;
    }
}
}
```

The ADC module can also be used to measure internal temperature of the microcontroller. We can perform this operation as in Listing 3.24. Here, we define the internal temperature sensor as int_temp(ADC_TEMP). As the code is executed, internal temperature of the STM32 microcontroller is obtained and printed in an infinite loop every 0.1 second.

Listing 3.24: Getting internal temperature of the microcontroller, the C code.

```
#include "mbed.h"

int main()
{
        AnalogIn int_temp(ADC_TEMP);

        while (true){
                int value = int_temp.read_u16();
                printf("Analog value read %u \n", (unsigned int)
                    value);
                wait_ms(100);
        }
}
```

3.3.3.4 DAC

We can also modify and use the DAC module via Mbed Studio. In the first example, we feed analog voltage to output from pin PA4. We provide the corresponding C code in Listing 3.25. Here, we define the analog output pin by AnalogOut dac(PA_4). The analog voltage to be fed to this pin is set by the function dac.write. This output can be observed by connecting a multimeter to pin PA4.

Listing 3.25: DAC output from pin PA4 using polling, the C code.

```
#include "mbed.h"

AnalogOut dac(PA_4);

int main()
```

3.3.4   C Control Systems Library

Since we will be using C language in digital control applications throughout the book, we introduce the corresponding C control systems library here. This is the C version of our MicroPython control systems library introduced in Section 3.2.4. Function names and entries are the same in both libraries. Hence, the reader can use both with minor modifications. To note here, C implementation will be slightly harder to use. More importantly, the C control systems library can only be used on the microcontroller.

We can include the C control systems library to our Mbed Studio project as follows. We should add the `Ccontrol.cpp` and `Ccontrol.h` files to our project. These files will be provided in the accompanying book website. Afterward, we should include the `Ccontrol.h` to our `main.cpp` file by the code snippet `#include "Ccontrol.h"`.

3.4   Application: Running the DC Motor

The aim in this application is to show how a system can be controlled by the STM32 microcontroller. We pick the DC motor as the system to explain the general setup for this purpose. As the basic operation, we will run the DC motor via C and Python codes. Therefore, we will first setup hardware components. Then, we will explain the procedure to be followed. Afterward, we will provide the C and Python codes used in the application. As a result, we will feed the necessary voltage value and direction information to the motor and observe how it works.

We provide block diagram of the general layout for this application in Figure 3.8. In this figure, $x[n]$ represents the system input. $x'[n]$ represents the signal generated by the microcontroller to be fed to the system. The amplifier represents the gain adjustment between $x[n]$ and $x'[n]$. The modules `System_Input` and `System_General` within the microcontroller stand for functions to control the system. We provide detailed explanation of these functions in Section 3.4.3.

Figure 3.8   General layout to control a system with microcontroller.

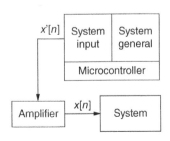

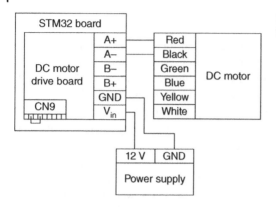

Figure 3.9 Hardware setup for the application.

3.4.1 Hardware Setup

To form the hardware setup for the application, we should place the DC motor driver expansion board onto the headers of the STM32 board. Then, the second upper pin of the connector CN9 of motor driver must be connected to the fourth upper pin of the connector CN9. Afterward, we should connect the positive and negative pins of the power supply module to the V_{in} and GND pins of the DC motor driver expansion board, respectively. The power supply should be set to 12 V. Finally, we should connect the DC motor's red and black cables to the motor driver expansion board's A+ (motor terminal) and A− (other motor terminal) inputs, respectively. We provide the hardware setup explaining these steps in Figure 3.9.

3.4.2 Procedure

As the reader forms the setup, the DC motor can be run either by the C or Python code given in the following section. Here, the DC motor input should be selected as 0, 3, 6, 9, or 12 and speed of the motor should be observed. Afterward, the motor input should be set as 18 and −18. The observed speed values corresponding to these inputs should be compared with the inputs 12 and 0, respectively. Finally, the reader should alter the direction of rotation for the DC motor by the menu for the C code and via its corresponding input value in the MicroPython code.

3.4.3 C Code for the System

In order to form the C code for the application, we should first form a new project under Mbed Studio. We should follow the steps explained in Section 3.3.1 for this purpose. As the project is generated, we should replace the content of the `main.cpp` file with the C code given in Listing 3.29. We should also place the `system_general.cpp`, `system_general.h`, `system_input.cpp`, system_input.h, system_output.cpp, system_output.h, system_controller.cpp, system_controller.h, menu.cpp, menu.h, matrix.cpp, matrix.h, Ccontrol.cpp, and Ccontrol.h files to the project. To note here, these files are provided in the accompanying book website.

Here, system_output.cpp, system_controller.cpp, matrix.cpp and Ccontrol.cpp are not used directly, but they are used by other files in the background.

Listing 3.29: The C code to be used in running the DC motor application.

```
#include "mbed.h"
#include "system_general.h"
#include "system_input.h"

Serial pc(USBTX, USBRX, 115200);
RawSerial device(PD_5, PD_6, 921600);
Ticker sampling_timer;
InterruptIn button(BUTTON1);
Timer debounce;
DigitalOut led(LED1);
PwmOut pwm1(PE_9);
PwmOut pwm2(PE_11);
DigitalOut pwm_enable(PF_15);
InterruptIn signal1(PA_0);
InterruptIn signal2(PB_9);
Timer encoder_timer;
AnalogIn ADCin(PA_3);
AnalogOut DACout(PA_4);

int main()
{
    user_button_init();
    system_input_init(1, 12/3.3);
    sampling_timer_init();
    menu_init();
    while (true) {
        system_loop();
    }
}
```

The menu.cpp file contains functions as a simple interface to be reached in the Mbed Studio terminal. Within this file, first the function menu_set_PWM_ frequency is called asking for the PWM signal frequency to run the DC motor. If the reader does not know this value, it should be initially set to the default value by entering 0. Afterward, the function menu_set_sampling_frequency is called and the program asks for the sampling frequency to be used in the system. Again, if this value is not known, it should be set to the default value suitable for our DC motor by entering 0. Then, the function menu_input_selection is called which forms a menu in the Mbed Studio terminal program. Here, the user can select one of the 12 input signals that can be applied to the DC motor input. In this chapter, we will only select the Constant Voltage input. We will explain other input signals, and their usage in detail in Chapter 4. As the `Constant Voltage` input is selected, the program asks for the voltage value to be fed to the DC motor. Afterward, the user should enter the direction of rotation for the DC motor. Then, the program informs the reader that the DC motor can be run by pressing the user button on the STM32 board.

As we check the `system_general.cpp` file, there are 11 functions there. Among them, `send_data_RT` and `send_variable` functions are related to sending data to PC and will be explained in Chapter 4. We briefly explain the remaining functions as follows: `isr_button(void)` is the interrupt callback function for the user button. It can be seen that this function is kept short. Only flags related to starting or stopping the system are set here. `isr_sampling_timer(void)` is the interrupt callback function for the sampling timer. This function is kept short again. Only the flag to start the periodic operations is set here. `sampling_timer_init(void)` function is used to select the sampling frequency which will be used by sampling timer through terminal. `user_button_init(void)` function is used to initialize the user button. `menu_init(void)` function is used to start the terminal interface on PC side. `system_loop(void)` function is executed in the main loop. All interrupt related processes are realized here. If `start_flag` is set, the system starts running. If `stop_flag` is set, the system stops. If `sample_flag` is set, system input is generated, system output is obtained, and desired data is sent to PC periodically. Here, the related flags should also be set through initialization. `start_function(void)` function is used to initialize the starting process inside the `system_loop` function when the `start_flag` is set. `stop_function(void)` function is used to initialize the stop process inside the `system_loop` function when the `stop_flag` is set. `select_signal(void)` function is used to produce system input sample according to the signal type inside the `system_loop` function.

As we check the `system_input.cpp` file, there are five functions in it. We briefly explain these functions as follows: `system_input_init(int type, float gain)` function is used to initialize system input block according to the `type` and `gain` inputs. `type` is used to select desired module to generate system input. Here, the user can select DAC, or two PWMs by choosing the `type` as 0, or 1, respectively. `gain` is used to set the constant gain value between system input voltage and microcontroller output voltage. `generate_system_input(float voltage)` function is used to generate microcontroller output according to the desired system input voltage, the gain difference between system input and microcontroller output, and selected module chosen by `type`. `convert_system_input_voltage(float voltage)` function is used to convert the desired system input voltage to microcontroller voltage level according to gain. set_duty_cycles(float voltage_mcu) function is used generate the desired PWM signals according to microcontroller voltage and desired motor rotation direction. set_DAC(float voltage_mcu) function is related to DAC to generate desired system input. This function will not be used in this application.

If we check the main.cpp file, it can be seen that user button is initialized first. Then, system input block is initialized by selecting two PWMs for DC motor input generation. The gain between motor input and microcontroller output is set as 12/3.3. Finally, sampling_timer_init and menu_init functions are called and the system_loop function is placed inside the infinite while loop.

After the input signal is selected with desired parameters, the system waits for the user to push the button to start the process. As the user button is pressed, the motor starts running. The green LED on the STM32 board also turns on to indicate this. There is also a notification on the terminal program telling that the motor is running, and the user button should be pressed to stop it. If the user presses the button again, the motor stops. The green LED on the STM32 board turns off. There is also a notification on the terminal program telling that the motor has stopped running and the user button should pressed to start running it again.

3.4.4 Python Code for the System

In order to form the Python code for the application, first we should replace the content of the main.py file with the Python code given in Listing 3.30 by following the steps explained in Section 3.2.2. We should also place the system_general.py, system_input.py, and mpcontrol.py files to flash of the microcontroller. To note here, these files are provided in the accompanying book website.

Listing 3.30: Python code to be used in running the DC motor application.

```
import system_general
import system_input

def main():
        system_general.button_init()
        system_general.PWM_frequency_select(0)
        system_general.sampling_timer_init(0)
        system_general.input_select(input_type='Constant Amplitude',
            amplitude=9)
        system_general.motor_direction_select(1)
        system_input.system_input_init(2, 3.63636363636)
        while True:
                system_general.system_loop()

main()
```

As we check the `system_general.py` file, there are 10 functions. Among them, `send_data_RT` and `send_input_output` functions are related to sending data to PC and will be explained in Chapter 4. We briefly explain the remaining functions next.

`isr_button(pin)` is the interrupt callback function for the user button. It can be seen that this function is kept short. Only flags related to starting or stopping the system are set here. `isr_sampling_timer(timer)` is the interrupt callback function for the sampling timer. This function is also kept short. Only the flag to start the periodic operations is set here. `sampling_timer_init(freq)` function is used to select the sampling frequency to be used in the system. If this value is not known, it should be set to the default value suitable for our DC motor by entering 0. `user_button_init()` function is used to initialize the user button. `PWM_frequency_select(freq)` function is used to select PWM signal frequency to run the DC motor. If the reader does not know this value, it should be initially set to the default value by entering 0. `motor_direction_select(direction)` function is used to select the rotation direction of DC motor. If `direction` is selected as 0, DC motor turns in counterclockwise direction. If it is selected as 1, DC motor turns in clockwise direction. `input_select(input_type, **kwargs)` function is used to select the desired system input with desired parameters. System input can be selected as one of the "Constant Amplitude," "Unit Pulse Signal," "Step Signal," "Ramp Signal," "Parabolic Signal," "Exponential Signal," "Sinusoidal Signal," "Damped Sinusoidal Signal," "Rectangular Signal," "Sum Of Sinusoids Signal," "Sweep Signal," or "Random Signal." Parameters of these signals can be seen in Section 4.1.3. In this application, we will only use the "Constant Amplitude" signal as input. `system_loop()` function is executed in the main loop. It can be seen that, all interrupt related processes are realized here. If `start_flag` is set, the system starts. If `stop_flag` is set, the system stops. If `sample_flag` is set, system input is generated, system output is obtained, and desired data is sent to PC periodically. Here, related flags should also be set through initialization.

As we check the `system_input.py` file, there are five functions. The function `system_input_init(type, gain)` is used to initialize the system input block according to `type` and `gain` inputs. `type` is used to select the desired module to generate system input. The user can select one of DAC, PWM, or two PWMs by choosing the `type` as 0, 1, or 2, respectively. `gain` is used to set the constant gain value between system input voltage and microcontroller output voltage. `convert_system_input_voltage(voltage, gain)` function is used to convert desired system input voltage to microcontroller voltage level according to `gain`. It can be seen that the microcontroller voltage is limited between 0 and 3.3 V. This means system input will also be limited according to the `gain` parameter. `set_duty_cycles(voltage_mcu, motor_dir)` function is used generate the desired PWM signal according to microcontroller voltage and motor rotation direction. set_DAC(voltage_mcu) function is related to DAC to generate the desired system input. This function will not be used in this application. generate_system_input(voltage, type, gain) function is used to generate the microcontroller output according to the desired system input voltage, the gain difference between system input and microcontroller output and selected module chosen by type.

If we check the main.py file, it can be seen that user button is initialized first. Then, PWM and sampling frequencies are selected using related functions. Afterward, system input type is selected as "Constant Amplitude," the voltage is set to 9 V, and motor rotation direction is set to clockwise. Then, the system input block is initialized by selecting two PWMs for DC motor input generation and the gain between motor input and microcontroller output is set as 12/3.3. Finally, the system_loop function is placed inside the infinite while loop. To note here, the mpcontrol library is not directly included in the main.py file. Instead, it is included by other files. Therefore, the reader should not think it as unnecessary.

As the user button is pressed, the motor starts running. The green LED on the STM32 board also turns on to indicate this. If the user presses the button again, the motor stops. The green LED on the STM32 board also turns off. The user should press the button again to start the motor.

3.4.5 Observing Outputs

In this application, we expect the reader to become familiar with the DC motor we will be using throughout the book. As the C or Python code is executed, the reader should observe that the motor rotates faster as the input voltage fed to it increases. The maximum speed can be observed when 12 V is applied. As 0 V is applied to the motor, it should stop. Since the gain parameter is set as 12/3.3 V, a value higher that 12 V will be saturated to this value. Likewise, a value entered smaller than 0 V will be saturated to 0 V. Finally, the reader should observe the effect of rotation direction input in the code.

3.5 Summary

Software is the most important element in digital control applications. We handled it in three steps as high, middle, and low in this chapter. As for high-level implementation, we picked the Python programming language. This language offers a user friendly environment on PC. As for the middle-level implementation, we picked MicroPython which is the modified form of Python for microcontrollers.

Finally, we picked the C language for the low-level operation. We provided sample codes on the usage of all three languages for various applications. To note here, these sample codes can be ported to other microcontrollers different than STM32. This is due to the abstraction property of MicroPython and Mbed Studio. However, the selected microcontroller should support these programming environments. Besides, it should handle the code size for the application. We also introduced Python, MicroPython, and C control systems libraries developed specifically for digital control applications. We will use them extensively in the following chapters. Finally, we provided an end of chapter application on running our DC motor by the STM32 microcontroller via C and Python codes. This will be the backbone of applications to be covered in the following chapters.

Problems

3.1 Provide a brief history of Python and analyze its recent popularity.

3.2 Download and install Python on your computer with all required libraries to be used throughout the book.

3.3 Write a program in Python language on PC
a. to calculate the first 10 elements of the function $\sin(\pi n/5)$ for $n = 0$ to 9. Store your results in an array.
b. Restructure your code such that the calculation is done within a function.
c. Add a code snippet to save your results to a file.

3.4 Write a program in Python language on PC
a. to calculate the first 10 elements of the Fibonacci series and store them in an array.
b. Restructure your code such that the calculation is done within a function.
c. Add a function to your code which takes the two calculated values in array form as input. If the array entry is even, it is replaced by one. Otherwise, it is replaced by zero. Return the calculated array as the output of your function.

3.5 What is the difference between Python and MicroPython?

3.6 Download and install MicroPython on your microcontroller. Make sure it works.

3.7 Repeat Problem 3.3 in MicroPython except part c. Observe the results on the command window.

3.8 Repeat Problem 3.4 in MicroPython. Observe the results on the command window.

3.9 Write a code in MicroPython such that the onboard red LED toggles every 0.1 second.

3.10 Expand Problem 3.9 such that the toggling operation starts when the "s" button is pressed on the PC keyboard. The toggling stops when the "d" button is pressed on the PC keyboard.

3.11 Open an account in http://mbed.org. Import and compile an example project related to the STM32 board we will be using throughout the book. Embed and execute the generated code on your board. Observe the results.

3.12 Download and install Mbed Studio on your computer. Compile, embed, and execute the available sample code there on the STM32 microcontroller to make sure that everything works as expected.

3.13 Repeat Problem 3.3 in C language except part c on Mbed Studio. Observe the results on the terminal window.

3.14 Repeat Problem 3.4 in C language on Mbed Studio. Observe the results on the terminal window.

3.15 Repeat Problem 3.9 in C language on Mbed Studio.

3.16 Repeat Problem 3.10 in C language on Mbed Studio.

4

Fundamentals of Digital Control

Digital control systems to be considered in the following chapters depend on digital signal representation. Therefore, this chapter starts with an introduction to digital signals both from theoretical and practical perspectives. Next, we will evaluate digital systems and their properties. Here, we will focus on linear and time-invariant (LTI) systems which form the fundamental structure in the following chapters. We will also explore z-transform and how it can be used in analyzing digital systems. We will follow a more practical approach throughout the chapter since our aim is applying the concepts introduced here to practical digital control problems. Therefore, we will provide Python and C codes whenever possible. We will direct the reader to references for in depth theoretical evaluation of digital control topics introduced here. As the end of chapter applications, we will first provide methods on handling offline signals on the STM32 microcontroller. Then, we will take the DC motor as the system and apply basic operations on it.

4.1 Digital Signals

We should acquire measurements from either a sensor or reference input to control a system. Here, we will get data by samples. We have seen how this can be done by the analog-to-digital converter (ADC) module of the STM32 microcontroller in Chapter 3. Although we will acquire data by samples, we will process them as an array. This forms the digital signal representation. We will start with the mathematical definition of such signals in this section. Then, we will focus on how they can be represented in Python or C code. Finally, we will consider standard digital signals to be used throughout the book.

Embedded Digital Control with Microcontrollers: Implementation with C and Python,
First Edition. Cem Ünsalan, Duygun E. Barkana, and H. Deniz Gürhan.
© 2021 The Institute of Electrical and Electronics Engineers, Inc. Published 2021 by John Wiley & Sons, Inc.
Companion website: www.wiley.com/go/Unsalan/Embedded_Digital_Control_with_Microcontrollers

4.1.1 Mathematical Definition

We can think of a digital signal as an array of numbers ordered by an index term. Hence, we will represent the digital signal as $x[n]$ throughout the book where the index n is integer. Based on this definition, $x[5]$ means that we are referring to the fifth element of the signal $x[n]$. Since we are dealing with an array of numbers as signal, the index term has to be integer. Otherwise, it would not make sense to reach a nonexistent array element. This array index can be positive or negative. However, we set the minimum index value to zero in practical applications since we assume that our signal starts at a certain time instant. Besides, we expect to have limited number of array elements since we will use them in code form in the following section.

The reader will often come up with the term "discrete-time" signal in literature. This also refers to an array of numbers which form a signal. The difference between the digital and discrete-time signals is in the value of array elements. To be more specific, digital signal entries can only be limited in range and value. In other words, the array elements can be of type `character`, `integer`, `float`, or `double` in Python or C languages. Hence, a digital signal can be processed by the code running on a digital system. On the other hand, the discrete-time signal does not have such limitation. Hence, its elements can take any value. We will show how to obtain the digital signal from its discrete-time counterpart in Chapter 5 while explaining the quantization operation. Although a discrete-time signal cannot be used in practical implementations, it simplifies life for mathematical derivations. Therefore, books focusing on the theory of digital signal processing or digital control almost always pick the discrete-time term instead of digital. We will follow the same path such that discrete-time signals will be used in mathematical derivations. However, we have to use the digital representation in practical implementations.

We will also use analog (or continuous-time) signals occasionally throughout the book, although our main focus is digital signals. The analog signal can be defined as a function in mathematical terms. Its usual representation is $x(t)$, where t is the index term. Here, the index t will be real such that it can take any value. Therefore, we can refer to a signal value such as $x(1.7509)$. Likewise, the analog signal at any index is real valued.

4.1.2 Representing Digital Signals in Code

Defining a digital signal in mathematical terms helps developing control methods. However, we will need the signal in actual implementation as well. Therefore, we will focus on this issue in Python and C languages next.

Before representing digital signals in Python and C languages, let us summarize some key concepts. First, the digital signal of interest will be defined as an array in Python and C languages. Second, the array (hence the signal) should be limited in time since we have limited space to store its entries in the microcontroller memory. Third, we will have index values always starting with zero since it is not possible to declare an array element starting with a negative value in Python or C. Fourth, we will have limited type definition options such as `character`, `integer`, `float`, and `double` for defining signal values.

4.1.2.1 Representation in Python

We can define an array in Python by the syntax in Listing 3.2. Python has a major advantage such that the user need not define the array type. The Python interpreter can adjust memory requirements by entries of the array.

We can reach array elements in Python by the syntax [·]. Hence, there is a natural resemblance between the mathematical definition of a digital signal and its Python representation. Let us provide an example on the usage of these on an actual signal. Assume that we have a digital signal with entries $-1, 0, 1, 2, 2/3, -3/4$, and 4.5. We can call this digital signal as x. Hence, it can be defined in Python as $x=$ `[-1, 0, 1, 2, 2/3, -3/4, 4.5]`. Moreover, we can reach a specific element of this digital signal as $x[\cdot]$. For example, we can reach $x[2]$ and print it on the command window as `print(x[2])`.

4.1.2.2 Representation in C

Defining an array in C language is not straightforward as in Python. The programmer should start with setting the type of array entries. This is a strict definition. Therefore, if the user defines the array to have `integer` entries at the beginning and by mistake a `float` number is assigned to the array, only the integer part of the entry can be kept. The array definition in C language has a syntax `type name [number of elements]`. A sample definition is as follows `int x[10]`. Here, we define an array x with 10 elements. Reaching a specific element within this array is the same as in Python. Hence, the reader can reach second element of the array by `x[2]`.

There is one important property called casting in C language which we will benefit from. Casting is used to change the type of a variable temporarily. Hence, if we have an integer array element and we want to process it in float form, what we do is cast it as float in the operation. The type of the element still stays as integer outside the operation. Let us give an example to this operation. Assume that the second element of the integer array x needs to be used in a float operation. We apply casting as `(float) x[2]`. Hence, it is treated as if it is a float only in that operation. Besides, value of the array element is kept as integer.

4.1.3 Standard Digital Signals

We will be receiving actual signals either from a sensor or reference input in practical applications. There are also predefined signals in order to explain the digital control concepts in mathematical terms. We will introduce them in this section.

4.1.3.1 Unit Pulse Signal

The first signal of interest is the unit pulse which can be used to model actual signals with very short duration. In mathematical terms, we can describe the unit pulse signal as having value one when its index is zero. Its value is zero for all other index terms. Hence, this signal can be defined as

$$\delta[n] = \begin{cases} 1 & n = 0 \\ 0 & \text{otherwise} \end{cases} \tag{4.1}$$

The unit pulse signal is defined within the MicroPython control systems library by the function `unit_pulse_signal(N)`, where N represents the total number of array elements to be returned. We can plot the unit pulse signal based on its Python representation. Letting N=49, we obtain the unit pulse signal as in Figure 4.1.

The unit pulse signal is defined within the C control systems library by the function `void unit_pulse_signal(int N, float signal[])`. Here, N represents the total number of array elements to be returned. `signal[]` is the array keeping unit pulse signal.

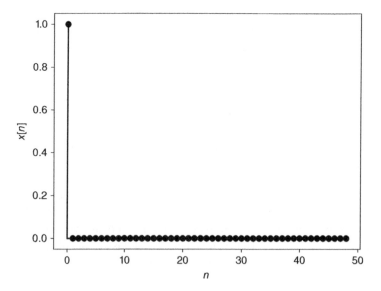

Figure 4.1 The unit pulse signal.

4.1.3.2 Step Signal

The second signal of interest is the step. This signal can be used to model an abrupt change to input of a digital system. In mathematical terms, the step signal takes two values as zero and A. If the index of the signal is greater than or equal to zero, then its value is A. Otherwise, its value is zero. Hence, this signal can be defined as

$$u[n] = \begin{cases} A & n = 0, 1, 2, \dots \\ 0 & \text{otherwise} \end{cases} \tag{4.2}$$

We can define the unit step signal when $A = 1$. We can also represent the unit step signal based on the unit pulse signal as $u[n] = \sum_{k=-\infty}^{n} \delta[k]$. This is summing (or integration) in digital domain. Hence, the operation is called integrator. Related to this, we can represent the unit pulse signal using two unit step signals as $\delta[n] = u[n] - u[n-1]$. This operation is called differentiator in digital domain. These representations will be of help in understanding digital system responses.

The step signal is defined within the MicroPython control systems library by the function `step_signal(N, Amp)`, where N represents the total number of array elements to be returned. `Amp` represents amplitude of the step signal. We should take this value to be one for the unit step signal. We can plot the unit step signal based on its Python representation. Letting `N=49` and `Amp=1`, we obtain the unit step signal as in Figure 4.2.

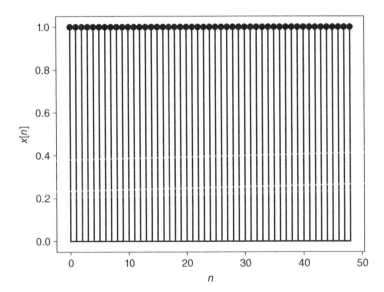

Figure 4.2 The unit step signal.

The step signal is defined within the C control systems library by the function `void step_signal(int N, float Amp, float signal[])`. Here, the parameters `N` and `Amp` are the same as in the corresponding MicroPython function. `signal[]` is the array keeping the step signal.

4.1.3.3 Ramp Signal

The third signal of interest is the ramp. This signal has an increasing value with respect to its index. Hence, it can be used to model actual signals with increasing value in time. One such example is the increasing speed of a motor.

Let us assume that the ramp signal starts from index value zero. Hence, we can represent it as

$$r[n] = \begin{cases} An & n = 1, 2, \ldots \\ 0 & \text{otherwise} \end{cases} \tag{4.3}$$

where A represents the amplitude term for the ramp signal. If we take $A = 1$, then we can obtain the unit ramp signal. Based on the definition in Eq. (4.3), we can represent the unit ramp signal as the cumulative sum of the unit step signal. In mathematical terms, it can be represented as $r[n] = \sum_{k=-\infty}^{n} u[k]$ as integration. Related to this, we can represent the unit step signal as the difference of successive unit ramp elements as $u[n] = r[n] - r[n-1]$ as differentiation.

The ramp signal is defined within the MicroPython control systems library by the function `ramp_signal(N, Amp)`, where `N` represents the total number of array elements to be returned. `Amp` represents amplitude of the signal. We should take this value to be one for the unit ramp signal. We can plot the ramp signal based on its Python representation. Letting `N=49` and `Amp=1`, we obtain the unit ramp signal as in Figure 4.3.

The ramp signal is defined within the C control systems library by the function `void ramp_signal(int N, float Amp, float signal[])`. Here, the parameters `N` and `Amp` are the same as in the corresponding MicroPython function. `signal[]` is the array keeping the ramp signal.

4.1.3.4 Parabolic Signal

The fourth signal of interest is parabolic. This signal has an increasing value with respect to its index. However, the increase is by square of the index value.

Let us assume that the parabolic signal starts from index value zero. We can represent it as

$$r[n] = \begin{cases} An^2 & n = 1, 2, \ldots \\ 0 & \text{otherwise} \end{cases} \tag{4.4}$$

where A is amplitude of the signal.

The parabolic signal is defined within the MicroPython control systems library by the function `parabolic_signal(N, Amp)`, where `N` represents the total number of array elements to be returned. `Amp` represents amplitude of the signal.

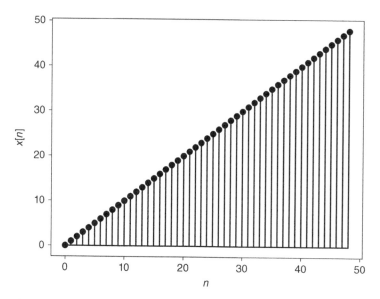

Figure 4.3 The unit ramp signal.

We can plot the parabolic signal based on its Python representation. Letting `N=49` and `Amp=1`, we obtain the parabolic signal as in Figure 4.4.

The parabolic signal is defined within the C control systems library by the function `void parabolic_signal(int N, float Amp, float signal[])`. Here, the parameters `N` and `Amp` are the same as in the corresponding MicroPython function. `signal[]` is the array keeping the parabolic signal.

4.1.3.5 Exponential Signal

The fifth signal of interest is the exponential which can be defined as

$$e[n] = \begin{cases} e^{an} & n = 0, 1, 2, \ldots \\ 0 & \text{otherwise} \end{cases} \tag{4.5}$$

where a is the decay parameter. Here, we limit the exponential signal such that it starts from index value zero.

The exponential signal can be used to model actual signals exponentially decreasing (or increasing) in time. This will be the case for some digital system responses. Therefore, we will see exponential signals frequently in the following chapters.

The exponential signal is defined within the MicroPython control systems library by the function `exponential_signal(N, a)`, where `N` represents the total number of array elements to be returned. Here, `a` represents decay parameter of the signal. We can plot the exponential signal based on its Python representation. Letting `N=49` and `a=-0.05`, we obtain the exponential signal as in Figure 4.5.

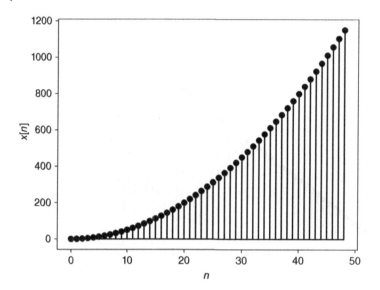

Figure 4.4 The parabolic signal.

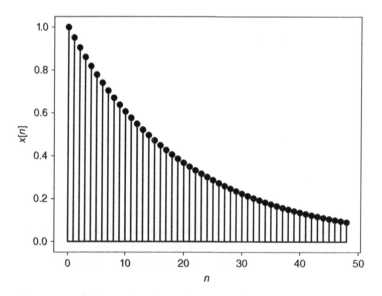

Figure 4.5 The exponential signal.

The exponential signal is defined within the C control systems library by the function `void exponential_signal(int N, float a, float signal[])` Here, the parameters `N` and `a` are the same as in the corresponding MicroPython function. `signal[]` is the array keeping the exponential signal.

4.1.3.6 Sinusoidal Signal

The sixth signal of interest is the sinusoidal, which can be defined as $s[n] = (A\sin(2\pi fn/f_s + \phi) + C)u[n]$. Here, A is the amplitude, f is the frequency, f_s is the sampling frequency, ϕ is the phase, and C is the offset term. To note here, we can also use the cosine function instead of sine here.

The sinusoidal signal is defined within the MicroPython control systems library by the function `sinusoidal_signal(N, Amp, Freq, Phase, Offset, Fsample, select)`. Here, `N` represents the total number of array elements to be returned. `Amp` represents amplitude of the signal. `Freq` stands for frequency of the signal. `Phase` is phase of the signal. `Offset` represents the offset value to be applied to the signal. `Fsample` is sampling frequency of the signal. This value will become clear after Chapter 5. Finally, `select` can be used to select between sine or cosine functions. If it is set as zero, the generated signal will be of type sine. Otherwise, the generated signal will be of type cosine. We can plot the sinusoidal signal based on its Python representation. Letting `N=49`, `Amp=1`, `Freq=2.5`, `Phase=0`, `Offset=0`, `Fsample=50`, and `select=0` we obtain the sinusoidal signal as in Figure 4.6.

The sinusoidal signal is defined within the C control systems library by the function `void sinusoidal_signal(int N, float Amp, float Freq, float Phase, float Offset, float Fsample, int select, float signal[])`. Here, the parameters `N`, `Amp`, `Freq`, `Phase`, `Offset`, `Fsample`, and `select` are the same as in the corresponding MicroPython function. `signal[]` is the array keeping the sinusoidal signal.

4.1.3.7 Damped Sinusoidal Signal

The seventh signal of interest is the damped sinusoidal which can be defined as multiplication of the exponential and sinusoidal signals. In other words, we can represent the damped sinusoidal signal as $s[n] = e^{an}(A\sin(2\pi fn/f_s + \phi) + C)u[n]$. Here, a is the decay parameter, A is the amplitude, f is the frequency, f_s is the sampling frequency, ϕ is the phase, and C is the offset term.

The damped sinusoidal signal is defined within the MicroPython control systems library by the function `damped_sinusoidal_signal(N, Amp_exp, Amp_sin, Freq, Phase, Offset, Fsample, select)`. Here, `N` represents the total number of array elements to be returned. `Amp_exp` is the decay parameter of the exponential part of the signal. `Amp_sin` represents amplitude of the sinusoidal part of the signal. `Freq` stands for frequency of the sinusoidal

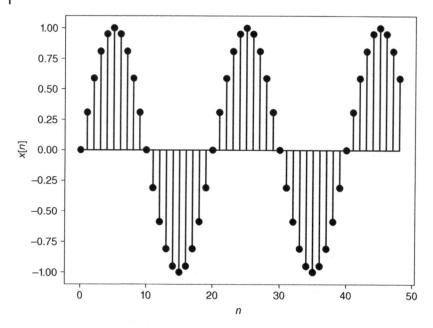

Figure 4.6 The sinusoidal signal.

part of the signal. Phase is phase of the sinusoidal part of the signal. Offset represents the offset value to be applied to the signal. Fsample is sampling frequency of the signal. This value will become clear after Chapter 5. Finally, select can be used to select between sine or cosine functions. If it is set as zero, the generated signal will be of type sine. Otherwise, the generated signal will be of type cosine. We can plot the damped sinusoidal signal based on its Python representation. Letting N=49, Amp_exp=-0.05, Amp=1, Freq=2.5, Phase=0, Offset=0, Fsample=50, and select=0 we obtain the damped sinusoidal signal as in Figure 4.7.

The damped sinusoidal signal is defined within the C control systems library by the function void damped_sinusoidal_signal(int N, float Amp_exp, float Amp_sin, float Freq, float Phase, float Offset, float Fsample, int select, float signal[]). Here, the parameters N, Amp_exp, Amp_sin, Freq, Phase, Offset, Fsample, and select are the same as in the corresponding MicroPython function. signal[] is the array keeping the damped sinusoidal signal.

4.1.3.8 Rectangular Signal

The eight signal of interest is the rectangular. This is a periodic signal defined within one period, T, as $A(u[n \bmod T] - u[n \bmod T - N_0]) + C$. Here, N_0 is the positive pulse duration. Duty cycle of the signal is $100 \times N_0/T$. C is the offset term.

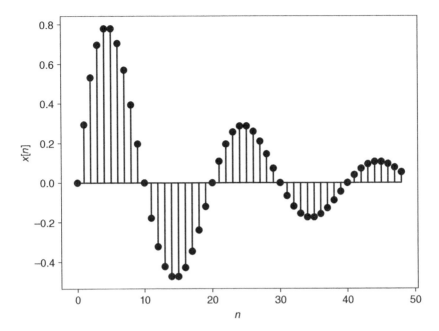

Figure 4.7 The damped sinusoidal signal.

This signal can also be named as pulse width modulation (PWM). Due to its excessive usage in driving DC motors, PWM signals are extremely important.

The rectangular signal is defined within the MicroPython control systems library by the function `rectangular_signal(N, Amp, Period, Duty, Offset)`. Here, `N` represents the total number of array elements to be returned. `Amp` is amplitude of the signal. `Period` stands for period of the signal. `Duty` is duty cycle of the signal. `Offset` represents the offset value to be applied to the signal. We can plot the rectangular signal based on its Python representation. Letting `N=49`, `Amp=1`, `Period=14`, `Duty=65`, and `Offset=0`, we obtain the rectangular signal as in Figure 4.8.

The rectangular signal is defined within the C control systems library by the function `void rectangular_signal(int N, float Amp, float Period, float Duty, float Offset, float signal[])`. Here, the parameters `N`, `Amp`, `Period`, `Duty`, and `Offset` are the same as in the corresponding MicroPython function. `signal[]` is the array keeping the rectangular signal.

4.1.3.9 Sum of Sinusoids Signal

The ninth signal of interest is the sum of sinusoids defined as $s[n] = \sum_{k=1}^{K}(A_k \sin(2\pi f_k n/f_s + \phi_k) + C_k)u[n]$. Here, A_k is the amplitude, f_k is the frequency, f_s is the sampling frequency, ϕ_k is the phase, and C_k is the offset term.

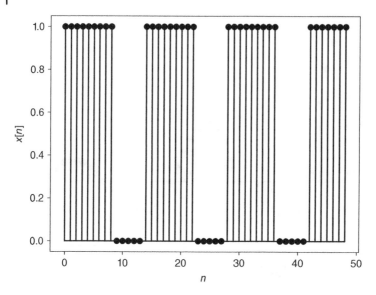

Figure 4.8 The rectangular signal.

To note here, we can also use cosine functions instead of sines here. As the name implies, the sum of sinusoids signal is composed of several sinusoidal signals. This type of signal will be of use in system identification applications to be introduced in Chapter 6.

The sum of sinusoids signal is defined within the MicroPython control systems library by the function `sum_of_sinusoids_signal(N, No_Sines, Amps, Freqs, Phases, Offsets, Fsample, select)`. Here, `N` represents the total number of array elements to be returned. `No_Sines` is the total number of sinusoids to be used in constructing the signal. `Amps` represents amplitude, `Freqs` stands for frequency, `Phases` is phase of the sinusoidal signals. `Offsets` represents the offset values to be applied to the sinusoidal signals. `Fsample` is sampling frequency of the sum of sinusoids signal. This value will become clear after Chapter 5. Finally, `select` value can be used to select between sine or cosine functions. If it is set as zero, the generated signal will be of type sum of sines. Otherwise, the generated signal will be of type sum of cosines. We can plot the sum of sinusoids signal based on its Python representation. Letting `N=49`, `Amps = [0.2, 0.2, 0.2, 0.2]`, `Freqs = [50/40, 50/30, 50/20, 50/10]`, `Phases = [0, 0, 0, 0]`, `Offsets = [0.25, 0.25, 0.25, 0.25]`, `Fsample=50`, and `select=0`, we obtain the sum of sinusoids signal as in Figure 4.9.

The sum of sinusoids signal is defined within the C control systems library by the function `void sum_of_sinusoids_signal(int N, int No_Sines,`

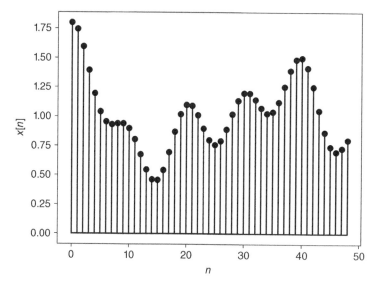

Figure 4.9 The sum of sinusoids signal.

`float Amps[], float Freqs[], float Phases[], float Off-sets[], float Fsample, int select, float signal[])`. Here, the parameters N, No_Sines, Amps, Freqs, Phases, Offsets, Fsample, and `select` are the same as in the corresponding MicroPython function. `signal[]` is the array keeping the sum of sinusoids signal.

4.1.3.10 Sweep Signal

The tenth signal of interest is the sweep which can be defined as $s[n] = (A \sin((2\pi f_1 n/f_s) + (2\pi(f_2 - f_1)n^2/(2f_s(d - 1)))) + C)u[n]$. Here, A is the amplitude, f_1 is the start frequency, f_2 is the stop frequency, f_s is the sampling frequency, d is the duration, and C is the offset term. Sweep signals are extensively used in applications in which a sinusoidal signal with varying frequency is required. The frequency range to be swept is set by f_1 and f_2.

The sweep signal is defined within the MicroPython control systems library by the function `sweep_signal(N, Amp, Freq_start, Freq_stop, Offset, Fsample, Duration)`. Here, N represents the total number of array elements to be returned. Amp represents amplitude of the signal. Freq_start and Freq_stop stand for the start and stop frequency values of the signal, respectively. Offset represents the offset value to be applied to the signal. Fsample is sampling frequency of the signal. This value will become clear after Chapter 5. Finally, Duration value can be used to select the duration of the signal. We can plot the sweep signal based on its Python representation.

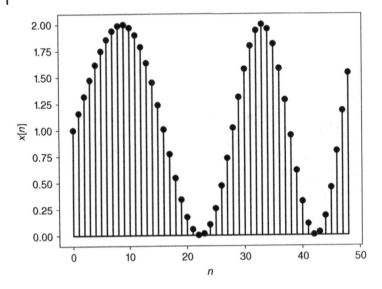

Figure 4.10 The sweep signal.

Letting N=49, Amp = 1, Freq_start = 50/40, Freq_stop = 50/10, Offset = 1, Fsample=50, and Duration=2, we obtain the sweep signal as in Figure 4.10.

The sweep signal is defined within the C control systems library by the function void sweep_signal(int N, float Amp, float Freq_start, float Freq_stop, float Offset, float Fsample, float duration, float signal[]). Here, the parameters N, Amp, Freq_start, Freq_stop, Offset, Fsample, and duration are the same as in the corresponding MicroPython function. signal[] is the array keeping the sweep signal.

4.1.3.11 Random Signal

The eleventh signal of interest is the random. This is a signal composed of random values. Randomness is generated by predefined functions in C and Python libraries. Random signals are extensively used in system identification to be considered in Chapter 6. Therefore, their generation and usage is extremely important.

The random signal is defined within the MicroPython control systems library by the function random_signal(N, Amp, Duration, Offset). Here, N represents the total number of array elements to be returned. Amp represents maximum amplitude of the random signal. Duration is the value such that the random sample is kept for that duration. Offset represents the offset value to be applied to the signal. We can plot the random signal based on its Python representation. Letting N=49, Amp = 1, Duration = 3, and Offset = 0 we obtain

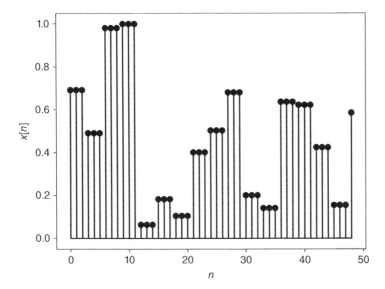

Figure 4.11 The random signal.

the random signal as in Figure 4.11. Please note that, the random signal will be different when generated in different times. Hence, the reader should not expect to observe the same signal when it is generated by his or her code.

The random signal is defined within the C control systems library by the function `void random_signal(int N, float Amp, int Duration,` `float Offset, float signal[])`. Here, the parameters `N`, `Amp`, `Duration`, and `Offset` are the same as in the corresponding MicroPython function. `signal[]` is the array keeping the random signal.

4.2 Digital Systems

Digital controllers to be developed in the following chapters will be in the form of systems. Besides, we will be controlling actual systems in implementation. Therefore, we will first provide mathematical definition of a digital system in this section. Then, we will show how it can be represented in code form which is the main topic of this book. Afterward, we will deal with system properties.

4.2.1 Mathematical Definition

We briefly introduced digital systems in Chapter 1. Here, we will provide their mathematical definition such that system properties can be derived. Mathematically, the digital system is an operator producing an output signal to a given input

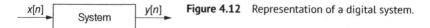

Figure 4.12 Representation of a digital system.

signal. In other words, the input signal is modified by the system to produce its output signal. The input signal can originate from a sensor, another system, or provided by the user (such as reference input). Output signal of the system can be used in another system or directly fed to an actuator. Such systems are called single input and single output (SISO). A system may take more than one input signal and produce more than one output signal. These systems are called multi-input multi-output (MIMO). Our main focus in this book will be SISO systems. Therefore, when we talk about a system it will always be of type SISO unless otherwise stated.

We can represent a digital system as in Figure 4.12 based on the digital signal representation introduced in the previous section. Here, $x[n]$ is the input signal fed to the system; $y[n]$ is the output signal obtained from the system. We will use this notation throughout the book.

4.2.2 Representing Digital Systems in Code

Analog systems can be constructed by physical elements such as resistor, capacitor, inductor, and operational amplifier. On the other hand, a digital system can be formed simply as a code block, written in Python, C, or another language. This code runs on a digital platform such as microcontroller. The only requirement here is representing mathematical definition of the digital system in code form. Let us consider digital system implementation in Python and C languages next.

4.2.2.1 Representation in Python

Python allows representing a digital system in code form. The most suitable structure to be used here is the function definition. Hence, the digital system can be described as a function with its input and output variables. Please refer to Section 3.1.5 on defining a function in Python. Let us pick a sample digital system as in the following example.

Example 4.1 *(Representing a sample digital system in Python)*
Assume that the input–output relation of a digital system is defined as $y[n] = Kx[n]$, where K is a constant. This system can be taken as the proportional controller which will be part of the more generalized PID control scheme to be explored in Section 8.1.3.

As we have the mathematical representation of the system, we can define it in Python as

```
def sample_system(x,K):
    y=K*x
    return y
```

As can be seen in this Python code, the system can be represented easily when its mathematical definition is available. Besides, the *K* parameter of the system can be modified in the Python code.

4.2.2.2 Representation in C
We can also use the C language to represent a digital system in code form. As in Python, we suggest using the function definition for this purpose. Let us reconsider the sample system introduced in the previous section and form its C function representation next.

Example 4.2 *(Representing a sample digital system in C)*
The sample digital system defined as $y[n] = Kx[n]$, where *K* is a constant can be represented in C code form as

```
void sample_system(float x, float K, float y){
    y=K*x;
}
```

As can be seen in this C code, the system can be represented easily when its mathematical definition is available. Besides, the *K* parameter of the system can be modified in the C code.

4.2.3 Digital System Properties

Knowing properties of a digital system helps us characterizing and implementing it. There are several system properties. Here, we will only focus on the stability, linearity, and time-invariance properties as the most important ones for digital control applications.

4.2.3.1 Stability
A system is stable if its output does not diverge to infinity for any given bounded input. To note here, bounded input means that input does not diverge to infinity. Hence, this stability definition is called bounded input bounded output (BIBO). There are also other stability definitions which we will not consider in this book. For more information on these, please see Ogata (1995). Next, we provide examples on stable and unstable systems.

Example 4.3   (*A stable system*)
The digital system $y[n] = 3x[n]$ is stable since its output is obtained by multiplying the input signal by a constant. Therefore, if the input signal is bounded, then there is no way the output can be unbounded.

Example 4.4   (*An unstable system*)
The digital system $y[n] = \sum_{k=-\infty}^{n} |x[k]|$ is unstable. Let us assume that the input to this system is $x[n] = u[n]$. The output with respect to for this input signal will reach infinity as the index n goes to infinity. Therefore, this system is unstable.

Checking stability of a system is not easy just by looking at its input–output relationship or by testing output of the system to several input signals as we have done in this section. There is a methodological way to check stability of LTI systems. Therefore, we will first introduce these next. Then, we will introduce stability check methods for such systems in Section 4.5.2.

4.2.3.2 Linearity
A system is called linear if it satisfies the homogeneity and additivity conditions. Let us say the input–output relationship of a system is represented by $y[n] = G\{x[n]\}$. The homogeneity condition can be represented as $G\{\gamma_1 x[n]\} = \gamma_1 G\{x[n]\}$, where γ_1 is constant. This condition indicates that when input of the linear system is multiplied by a constant, then the obtained output from the system should also be multiplied by the same value. The additivity condition can be represented as $G\{\gamma_1 x_1[n] + \gamma_2 x_2[n]\} = \gamma_1 G\{x_1[n]\} + \gamma_2 G\{x_2[n]\}$, where γ_1 and γ_2 are constants. The additivity condition indicates that a linear system can process an input signal as a whole or by parts such that the output will be the same for both cases. If a system satisfies both homogeneity and additivity conditions, then it is called linear. Otherwise, it is called nonlinear. Next, we provide examples on linear and nonlinear systems.

Example 4.5   (*A linear system*)
The digital system $y[n] = 3x[n]$ is linear since we can check the homogeneity and additivity conditions are satisfied for any γ_1 and γ_2 values.

Example 4.6   (*A nonlinear system*)
The digital system with the input–output relationship given in Eq. (4.6) is nonlinear.

$$y[n] = \begin{cases} 3x[n] & |x[n]| < 1 \\ 3 & x[n] > 1 \\ -3 & x[n] < -1 \end{cases} \tag{4.6}$$

One way to see this result is feeding a unit-step signal to the system and feeding a step signal with amplitude four. Output obtained from both inputs will be different besides their amplitude values. Hence, we can deduce that the system of interest is nonlinear.

4.2.3.3 Time-Invariance

Time-invariance property tells us that the system characteristics do not change with time. Hence, the system does not have different responses to the same input applied in different times. This definition also leads to a method to test the time-invariance property of a given system.

Let us assume that we feed an input signal $x[n]$ to a digital system and obtain its output as $y[n]$. If this system is time-invariant, then a time shift in the input signal should be observed in output as well. In other words, when we feed a shifted input signal as $x[n - n_0]$, we should get $y[n - n_0]$ for a time-invariant digital system. On the other hand, if a time shift in the input signal does not produce the same amount of shift in the output signal, then the system is time-varying. Next, we provide examples on time-invariant and time-varying systems.

Example 4.7 *(A time-invariant system)*
The digital system $y[n] = 3x[n]$ is time-invariant. As in the definition of time-invariance, a shift in the input signal such as $x[n - n_0]$ leads to the same shift in the output as $y[n - n_0]$. Therefore, it is time-invariant.

Example 4.8 *(A time-varying system)*
The digital system $y[n] = x[2n]$ is time-varying. As can be seen in the definition of the system, when we shift the input signal by n_0, the corresponding output will be shifted by $2n_0$ as $y[n - 2n_0] = x[2n - 2n_0]$. As can be seen here, the shift in the output is not the same as the shift in input. Therefore, this is a time-varying system.

4.3 Linear and Time-Invariant Systems

Digital control applications cover wide range of systems including nonlinear and time-varying ones. Among these, LTI systems deserve special consideration. Therefore, we will focus on them in this section. We will start with their mathematical definition. Afterward, we will associate constant-coefficient difference equations and LTI systems. We will also evaluate representing LTI systems in code form. Finally, we will look at the ways of connecting simple LTI systems to form more complex systems.

4.3.1 Mathematical Definition

By definition, an LTI system satisfies both linearity and time-invariance properties. These lead to an important property of an LTI system in terms of its input–output

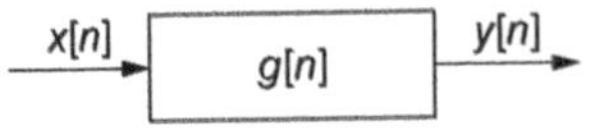

$x[n]$ → | $g[n]$ | → $y[n]$

Figure 4.13 Representation of an LTI system with impulse response $g[n]$.

relationship. More specifically, input–output relation of an LTI system can be represented by convolution sum with the formula

$$y[n] = \sum_{k=-\infty}^{\infty} x[k]g[n-k] \tag{4.7}$$

The shorthand representation of the convolution operation is $y[n] = x[n] * g[n]$. Here, $g[n]$ is impulse response of the system. To note here, impulse response of a system is the output obtained when the system input is the unit pulse signal, $\delta[n]$. We should note here that the term "impulse response" originates from the continuous-time systems and impulse signal $\delta(t)$. Since the impulse response term became standard, it is also used in digital systems as well. We also follow the same standard in this book.

As can be seen in Eq. (4.7), if we know $g[n]$, then we can calculate output of the system for any input. This makes LTI systems popular among others. We can represent an LTI system with impulse response $g[n]$ in schematic form as in Figure 4.13. z-Transform of $g[n]$ is called transfer function of the system. We will see how to obtain it in Section 4.5.

4.3.2 LTI Systems and Constant-Coefficient Difference Equations

The input–output relationship of an LTI system can also be represented by its corresponding constant-coefficient difference equation. This difference equation will have the general form of

$$y[n] + a_1 y[n-1] + \cdots + a_{L-1} y[n-L+1] = b_0 x[n] + \cdots + b_{K-1} x[n-K+1] \tag{4.8}$$

We will call b_k for $k \in [0, K-1]$ as numerator coefficients and a_l for $l \in [1, L-1]$ as denominator coefficients in this equation. This naming convention will become evident when we cover z-transform in Section 4.4.

We can leave $y[n]$ alone in Eq. (4.8) and rewrite it in closed form as

$$y[n] = \sum_{k=0}^{K-1} b_k x[n-k] - \sum_{l=1}^{L-1} a_l y[n-l] \tag{4.9}$$

There are two sum terms in Eq. (4.9). The first one only takes current and past input signal values. This sum is called the feedforward term. The second sum only takes past output signal values. This sum is called the feedback term.

4.3.3 Representing LTI Systems in Code

The representation in Eq. (4.9) allows us to implement the LTI system in code form. To do so, we should store the past output and input values as well as the present input signal value. Therefore, we will have a recursive implementation. Since LTI systems are important, their input and output relation is specifically represented in Python, MicroPython, and C control systems libraries. Let us consider them next.

4.3.3.1 MicroPython Control Systems Library Usage

We can create an LTI system by using the function g=tf(num,den,Ts) in the MicroPython control systems library. Here, num represents the numerator coefficient array. den represents the denominator coefficient array. Ts is the sampling period for digital systems. When it is not provided, the system is taken as continuous-time.

As we form the LTI system, we can use the available step(g, T=None) function in the MicroPython control systems library to calculate its step response. Likewise, we can use the function lsim(g, x, T=None) in the same library to calculate response of the system to the given input signal x. Let us provide an example on the usage of these structures.

Example 4.9 *(Representing an LTI system by the MicroPython control systems library)*

Assume that we have a digital system represented by its difference equation $y[n] = 0.9797y[n-1] + 0.2128x[n-1]$. In fact, this digital system is the first-order representation of our DC motor introduced in Chapter 2. The numerator and denumerator coefficients of the digital system can be represented as num=[0.2128] and den=[1, -0.9797] respectively. We can form the corresponding transfer function as g = mpcontrolPC.tf(num, den, Ts), where Ts = 0.0005.

We can use the constructed transfer function to observe its step response. To do so, we should form the Python code as in Listing 4.1.

Listing 4.1: Representing an LTI system by the MicroPython control systems library.

```
import mpcontrol
import mpcontrolPC

# System
Ts = 0.0005
num = [0.2128]
den = [1, -0.9797]
g = mpcontrolPC.tf(num, den, Ts)

N = 400
#Step input signal
x = mpcontrol.step_signal(N, 1)

#Calculate the output signal
y = mpcontrol.lsim(g, x)
```

We can plot the obtained output signal using the matplotlib library. As a result, we will obtain the output signal in Figure 4.14.

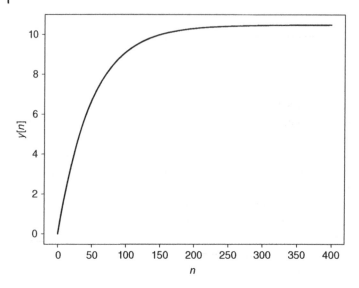

Figure 4.14 Unit step response of the DC motor.

4.3.3.2 C Control Systems Library Usage

We can also represent an LTI system by the help of the C control systems library. To do so, we should first define a transfer function prototype using the predefined structure `tf_struct`. Then, we can use the function `void create_tf(tf_struct *tf, float *num, float *den, int nr_num_coeffs, int nr_den_coeffs, float ts)` under the C control systems library to create the transfer function. Here, `tf` is the transfer function structure to be generated. `num` and `den` represent the numerator and denominator polynomials of the LTI system to be constructed, respectively. `nr_num_coeffs`, and `nr_den_coeffs` stand for the number of numerator and denominator coefficients. `ts` is the sampling period for digital systems.

As we construct the transfer function by the above structure, we can obtain its output `y` for a given input `x` using the function `void lsim(tf_struct *tf, int N, float x[], float y[])`. Let us provide an example to the usage of these structures next.

Example 4.10 (*Representing an LTI system by the C control systems library*)
Let us reconsider the digital system with the difference equation in Example 4.9. We can form the transfer function and obtain output of the system to step input with the C code in Listing 4.2. We will see how to plot the obtained signal in Section 4.6.

Listing 4.2: Representing an LTI system by the C control systems library.

```c
#include "mbed.h"
#include "Ccontrol.h"

#define N 400

RawSerial device(PD_5, PD_6, 921600);

// System
tf_struct g;
float Ts = 0.0005;
float num[1] = {0.2128};
float den[2] = {1, -0.9797};
float x[N], y[N];

int main()
{
create_tf(&g, num, den, sizeof(num)/sizeof(num[0]),sizeof(den)/
    sizeof(den[0]), Ts);

//Step input signal
step_signal(N, 1, x);

//Calculate the output signal
lsim(&g, N, x, y);

while(true){}
}
```

4.3.3.3 Python Control Systems Library Usage

We can create an LTI system by using the function tf(num,den,[,dt]) in the Python control systems library. Here, num represents the numerator coefficient array. den represents the denominator coefficient array. dt stands for the sampling time. As we form the system, we can use the available forced_response, impulse_response, initial_response, and step_response functions in the Python control systems library to calculate the response of a given system to a specific input signal. Let us provide an example on the usage of these functions.

Example 4.11 *(Representing an LTI system by Python control systems library)*

Let us reconsider the digital system in Example 4.9. We can use the Python control system library to represent this system. The numerator and denumerator coefficients of the digital system can be represented as num=[0.2128] and den=[1, -0.9797], respectively. We can set dt as 0.0005. Afterward, we can form the transfer function for the digital system as g=tf(num,den,1). We can use the constructed transfer function to observe its step response. To do so, we should form the Python code as in Listing 4.3.

Listing 4.3: Representing an LTI system by the Python control systems library.

```
import control

N = 400

# System
Ts=0.0005
num = [0.2128]
den = [1, -0.9797]
t = [n*Ts for n in range(N)]
g = control.TransferFunction(num, den, Ts)

#Calculate step response of the system
n,y=control.step_response(g, T=t)
```

There are functions in the MicroPython control systems library to convert LTI system representations between MicroPython and Python control systems library formats. To do so, the reader should use the function mptf_to_ tf(tf_mpcontrol) to convert the system representation from MicroPython to Python format. The function tf_to_mptf(tf_mpcontrol) performs the reverse operation. These two functions allow us to use both control systems libraries interchangeably in the following chapters. Hence, we can benefit from both in applications. Next, we provide an example on the usage of these conversion functions.

Example 4.12 *(Converting LTI system representations between Python and MicroPython control systems libraries)*
Let us reconsider the digital system in Example 4.9. We represented this digital system both in MicroPython and Python control systems libraries in previous examples. The Python code in Listing 4.4 can be used to convert these representations between both library formats.

Listing 4.4: Converting LTI system representations between Python and MicroPython control systems library formats.

```
import control
import mpcontrolPC

#System definition in MicroPython control systems library
Ts = 0.0005
num = [0.2128]
den = [1, -0.9797]
gmp = mpcontrolPC.tf(num, den, Ts)

print(gmp)

# System definition in Python control systems library
Ts = 0.0005
num = [0.2128]
den = [1, -0.9797]
gc = control.TransferFunction(num, den, Ts)
```

```
print(gc)

gmp2c = mpcontrolPC.mptf_to_tf(gmp)

print(gmp2c)

gc2mp = mpcontrolPC.tf_to_mptf(gc)

print(gc2mp)
```

4.3.4 Connecting LTI Systems

Simple systems can be connected to form more complex systems. There are three standard connection options as series (cascade), parallel, and feedback. Although these connections are not specific to LTI systems, they have a major advantage in this form. Connecting two or more LTI systems also results in another LTI system. Hence, the resultant system can be analyzed and implemented in the same way as an LTI system.

4.3.4.1 Series Connection

There may be cases in which a signal is fed to the first system, $g_1[n]$, as input. Output of this first system can be fed to the second system, $g_2[n]$, as input. Such a connection between the two systems is called series (cascade). We provide schematic representation of this setup in Figure 4.15. The equivalent impulse response between input $x[n]$ and output $y[n]$ will be as $g[n] = g_1[n] * g_2[n]$.

We can use the MicroPython control systems library to obtain the equivalent transfer function of two serially connected LTI systems using the function `series(tf1, tf2)`, where `tf1` and `tf2` represent the two transfer functions to be connected in serial form. Let us provide an example to this usage.

Example 4.13 *(Series connection of two LTI systems by the MicroPython control systems library)*
In order to connect two LTI systems in serial form, we should first construct them using the MicroPython control systems library. Then, we can use the function `series()` to connect them. The Python code in Listing 4.5 shows these steps.

Figure 4.15 Series connection of two LTI systems.

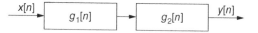

Listing 4.5: Series connection of two LTI systems by the MicroPython control systems library.

```
import mpcontrolPC

# Systems
Ts = 1
num1 = [1]
den1 = [1, -0.5]
g1 = mpcontrolPC.tf(num1, den1, Ts)

num2 = [1]
den2 = [1, 0.5]
g2 = mpcontrolPC.tf(num2, den2, Ts)

#Form the series connection
g3 = mpcontrolPC.series(g1, g2)
```

Likewise, we can use the Python control systems library to obtain the equivalent impulse response of serially connected systems. The function `series(sys1, *sysn)` within the library can be used to construct the series connection of *n* systems represented by their transfer function. The reader can check the usage of this function through the Python control systems library website.

4.3.4.2 Parallel Connection

There may be cases in which a signal is fed to two systems at once and their outputs are summed. Such a connection between the two systems is called parallel. We provide schematic representation of this setup in Figure 4.16. The equivalent impulse response between input $x[n]$ and output $y[n]$ for this connection will be as $g[n] = g_1[n] + g_2[n]$, where $g_1[n]$ and $g_2[n]$ are the impulse response of the first and second systems, respectively.

We can use the MicroPython control systems library to obtain the equivalent transfer function of two LTI systems connected in parallel form using the function `parallel(tf1, tf2)`. Here, `tf1` and `tf2` represent the two transfer functions to be connected in parallel form. Let us provide an example to this connection.

Example 4.14 *(Parallel connection of two LTI systems by the MicroPython control systems library)*

In order to connect two LTI systems in parallel form, we should first construct them using the MicroPython control systems library. Then, we can use the

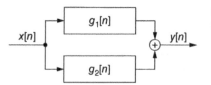

Figure 4.16 Parallel connection of two LTI systems.

function `parallel()` to connect them. The Python code in Listing 4.6 shows these steps.

Listing 4.6: Parallel connection of two LTI systems by the MicroPython control systems library.

```
import mpcontrolPC

# Systems
Ts = 1
num1 = [1]
den1 = [1, -0.5]
g1 = mpcontrolPC.tf(num1, den1, Ts)

num2 = [1]
den2 = [1, 0.5]
g2 = mpcontrolPC.tf(num2, den2, Ts)

#Form the parallel connection
g3 = mpcontrolPC.parallel(g1, g2)
```

We can also use the Python control systems library to obtain the equivalent transfer function of parallel connected systems. The function `parallel(sys1, *sysn)` can be used for this purpose. The reader can check the usage of this function through the Python control systems library website.

4.3.4.3 Feedback Connection

There may be cases in which a signal is fed to one system and its output is processed by the second system. When output of the second system is fed back to input of the first system again, a feedback loop is formed. We provide schematic representation of this setup in Figure 4.17. Here, $g_1[n]$ and $g_2[n]$ are the impulse response of the first and second systems, respectively. We can obtain the equivalent impulse response between input $x[n]$ and output $y[n]$ for this connection after introducing the z-transform in Section 4.4.

We can use MicroPython control systems library to obtain the equivalent transfer function of two LTI systems connected in feedback form using the function `feedback(tf1, tf2, sign)`, where `tf1` and `tf2` represent the two transfer functions to be connected in feedback form. `sign` stands for the sign of the feedback connection. If it is set as −1, there will be a negative feedback. This is the default value. Hence, when the reader does not enter any value to `sign`, the feedback connection is set as negative. If `sign` is set to 1, then a positive feedback loop will be formed. Let us provide such an example.

Figure 4.17 Feedback connection of two LTI systems.

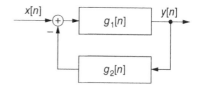

Example 4.15 (*Feedback connection of two LTI systems by the MicroPython control systems library*)
In order to connect two LTI systems in feedback form, we should first construct them using the MicroPython control systems library. Then, we can use the function feedback() to connect them. The Python code in Listing 4.7 shows these steps.

Listing 4.7: Feedback connection of two LTI systems by the MicroPython control systems library.

```
import mpcontrolPC

# Systems
Ts = 1
num1 = [1]
den1 = [1, -0.5]
g1 = mpcontrolPC.tf(num1, den1, Ts)

num2 = [1]
den2 = [1, 0.5]
g2 = mpcontrolPC.tf(num2, den2, Ts)

#Form the feedback connection
g3 = mpcontrolPC.feedback(g1, g2)
```

We can use the Python control systems library to obtain the equivalent impulse response of two systems forming a feedback connection. The function feedback(sys1, [sys2, sign]) can be used for this purpose. This function also allows the reader to add sign of the connection as positive or negative. The reader can check the usage of this function through the Python control systems library website.

4.4 The *z*-Transform and Its Inverse

We represented digital signals and systems in time domain up to this point. Although we can learn a lot from these in time domain, this is not sufficient. Therefore, we may need to represent digital signals and systems in complex domain as well. The method for switching between time and complex domains is the *z*- and inverse *z*-transform. We will focus on these transforms in this section. We will also calculate them via Python code on PC.

4.4.1 Definition of the *z*-Transform

The *z*-transform of a digital signal $x[n]$ can be calculated as $X(z) = \sum_{n=-\infty}^{\infty} x[n]z^{-n}$, where z is the complex variable represented in polar form as $z = re^{j\omega}$. Here, r is the radius term which cannot be negative. ω is the angle (phase) term having value between $[0, 2\pi]$ radians. As can be seen in the definition of z-transform, it is in fact the infinite sum of $x[n]$ multiplied by z^{-n} in terms of index n. Hence, the result will not depend on the index value. This infinite sum will be valid only if it converges. z values satisfying this convergence constraint in complex plane are called the region of convergence (ROC). Thus, any z-transform should have its accompanying ROC.

We can calculate the z-transform of known discrete-time signals. Let us start with the unit pulse signal.

Example 4.16 *(z-Transform of the unit pulse signal)*
The unit pulse signal has the z-transform as $X(z) = \sum_{n=-\infty}^{\infty} \delta[n]z^{-n} = 1$. As can be seen here, z-transform of the unit pulse signal is one. ROC for this case is the overall complex plane since the z-transform is independent of z.

Example 4.17 *(z-Transform of the unit step signal)*
z-Transform of the unit step signal can be calculated as

$$X(z) = \sum_{n=-\infty}^{\infty} u[n]z^{-n}$$

$$= \sum_{n=0}^{\infty} z^{-n}$$

$$= \frac{1}{1-z^{-1}} \tag{4.10}$$

ROC for this z-transform is $|z| > 1$ since the infinite sum in z-transform converges only under this constraint.

Example 4.18 *(z-Transform of the generalized exponential signal)*
z-Transform of the generalized exponential signal $x[n] = a^n u[n]$ can be calculated as

$$X(z) = \sum_{n=-\infty}^{\infty} a^n u[n]z^{-n}$$

$$= \sum_{n=0}^{\infty} (a^{-1}z)^{-n}$$

$$= \frac{1}{1-az^{-1}} \tag{4.11}$$

Here, ROC is $|z| > |a|$ since the infinite sum in z-transform converges only under this constraint.

4.4.2 Calculating the *z*-Transform in Python

We can use Python and its `sympy` library (with url https://www.sympy.org) on PC to calculate the *z*-transform of a signal. The `sympy` library is introduced to perform symbolic calculations. To do so, the reader should define symbols by the function `symbols`. Afterward, operations can be done on these symbols. The result can be printed by the function `pprint`. Below, we provide such an example.

Example 4.19 *(Calculating the z-transform of a signal in Python)*
We can calculate the *z*-transform of $x[n] = 0.3^n u[n]$ in Python. We can use the Python code in Listing 4.8 on PC for this purpose. As can be seen here, obtaining the *z*-transform is straightforward in Python. The obtained result also indicates the ROC for the infinite sum to converge.

Listing 4.8: Calculating the *z*-transform of a signal in Python.

```
from sympy import *

n = symbols('n')
z = symbols('z')

Tr = symbols('Tr')

# z-transform
Tr=summation(z**-(0.3*n), (n, 0, oo))
pprint(Tr)
```

4.4.3 Definition of the Inverse *z*-Transform

As we transform a time-domain signal to complex domain using *z*-transform, we may also need the reverse operation. This can be done by the inverse *z*-transform, which converts a complex domain representation to time-domain. The formula for this operation is

$$x[n] = \frac{1}{2\pi j} \oint_C X(z) z^{n-1} \, dz \tag{4.12}$$

where C represents the closed contour within ROC of the *z*-transform. Unfortunately, the integral in Eq. (4.12) cannot be calculated via standard Riemann integration methods. Therefore, complex integration methods should be used to calculate the result. This is not easy. However, we can benefit from Python on PC to calculate the inverse *z*-transform as well. Next, we will focus on this issue.

4.4.4 Calculating the Inverse *z*-Transform in Python

We can use the `sympy` library in Python to calculate the inverse *z*-transform as well. To do so, we should use the `series` function defined within this library. Below, we provide such an example.

Example 4.20 *(Calculating the inverse z-transform of a signal in Python)*
We can calculate the inverse z-transform of $X(z) = \frac{1}{1-z^{-1}}$ in Python. To do so, we can use the Python code in Listing 4.9 on PC.

Listing 4.9: Calculating the inverse z-transform of a signal in Python.

```
from sympy import *

z = symbols('z')
ITr = symbols('ITr')

# inverse z-transform
ITr=series(1/(1+z**(-1)),z**(-1))

print(ITr)
```

As the Python code in Listing 4.9 is executed on PC, we will obtain the result as `-1/z**5 + z**(-4) - 1/z**3 + z**(-2) - 1/z + 1 + O(z** (-6), (z, oo))`. This form can be converted to standard z-transformation representation to construct $x[n]$ from it.

As can be seen here, obtaining the inverse z-transform is straightforward in Python on PC. To note here, the closed form representation is not provided as the final result. The reader should remember that we are calculating the inverse z-transform as a series operation. That is why such a result is obtained.

4.5 The *z*-Transform and LTI Systems

z-Transform provides two useful tools for analyzing LTI systems. The first one associates the constant-coefficient difference equation and impulse response of the system. The second tool allows analyzing stability of an LTI system in complex domain. We will introduce these next.

4.5.1 Associating Difference Equation and Impulse Response of an LTI System

We can use z-transform to represent the constant-coefficient difference equation of an LTI system in complex domain. This leads to obtaining impulse response of the system in time domain. Let us consider an LTI system with impulse response $g[n]$. We know that input and output of this system can be represented as $y[n] = g[n] * x[n]$ from the convolution sum formula. If we take the z-transform of this convolution sum, we will obtain $Y(z) = G(z)X(z)$ where $Y(z)$, $G(z)$, and $X(z)$ are the z-transform of $y[n], g[n],$ and $x[n]$, respectively (Oppenheim et al., 1997). Here, $G(z)$ is transfer function of the system, and we can write it as

$$G(z) = \frac{Y(z)}{X(z)} \tag{4.13}$$

We know that the same LTI system can also be represented by its constant-coefficient difference equation from Eq. (4.8). Taking the z-transform of this equation, we will have

$$X(z) \sum_{k=0}^{K-1} b_k z^{-k} = Y(z) \sum_{l=0}^{L-1} a_l z^{-l} \tag{4.14}$$

where $a_0 = 1$. We will call b_k for $k \in [0, K-1]$ as numerator coefficients and a_l for $l \in [0, L-1]$ as denominator coefficients.

In Eq. (4.14), the difference equation in time domain became an algebraic equation in complex domain. We can arrange this equation to obtain

$$G(z) = \frac{Y(z)}{X(z)} = \frac{\sum_{k=0}^{K-1} b_k z^{-k}}{\sum_{l=0}^{L-1} a_l z^{-l}} \tag{4.15}$$

Hence, $G(z)$ becomes the ratio of two polynomials for an LTI system. The numerator polynomial holds the coefficient of $x[n]$ terms in the constant-coefficient difference equation. The denominator polynomial holds the coefficient of $y[n]$ terms.

One can obtain the time domain representation of impulse response of an LTI system by taking the inverse z-transform of Eq. (4.15). As can be seen here, z-transform also allows us to derive the impulse response of an LTI system from its constant-coefficient difference equation. We provide such an example next.

Example 4.21 *(Obtaining impulse response of a system from its difference equation)*

Let us reconsider the system in Example 4.9 with the difference equation $y[n] = 0.9797y[n-1] + 0.2128x[n-1]$. Let us represent this difference equation in z-domain. We will have

$$Y(z) = 0.9797z^{-1}Y(z) + 0.2128X(z) \tag{4.16}$$

Rearranging the terms in Eq. (4.16), we can obtain the corresponding transfer function as

$$G(z) = \frac{Y(z)}{X(z)} = \frac{0.2128}{z - 0.9797} \tag{4.17}$$

Since $G(z)$ is available, its time domain representation can be obtained as explained in Section 4.4.3. To do so, let us use the Python code in Listing 4.10 on PC.

Listing 4.10: Obtaining the impulse response of a system from its difference equation.

```
from sympy import *

z = symbols('z')
ITr = symbols('ITr')
```

```
# inverse z-transform
ITr = symbols('ITr')
ITr=series(0.2128*z**(-1)/(1-0.9797*z**(-1)),z**(-1))

print(ITr)
```

As the Python code in Listing 4.10 is executed on PC, we will obtain
`0.1961/z**5 + 0.2001/z**4 + 0.2043/z**3 + 0.2085/z**2 + 0.2128/z + O(z**(-6)), (z, oo))`. Hence, we can construct at least the first five elements of $g[n]$ as $g[n] = 0.2128\delta[n-1] + 0.2085\delta[n-2] + 0.2043\delta[n-3] + 0.2001\delta[n-4] + 0.1961\delta[n-5]$.

z-Transform can also be used to construct the constant-coefficient difference equation of an LTI system from its impulse response. To do so, the first step is taking the z-transform of $g[n]$ to obtain $G(z)$. Then, $G(z)$ can be used in Eq. (4.15) such that an algebraic equation can be formed between the $X(z)$ and $Y(z)$ terms as in Eq. (4.13). Finally, applying the inverse z-transform to this equation leads to the constant-coefficient difference equation representation of the LTI system in time domain.

4.5.2 Stability Analysis of an LTI System using z-Transform

z-Transform can be used to analyze stability of an LTI system. To do so, we should first define what zero and pole of a system means. As can be seen in Eq. (4.15), $G(z)$ can be represented as the ratio of two polynomials. Roots of these polynomials can be extracted. Hence, Eq. (4.15) can be written as

$$G(z) = \frac{(z - z_0)(z - z_1) \cdots (z - z_K)}{(z - p_0)(z - p_1) \cdots (z - p_L)} \qquad (4.18)$$

where $z_0 \cdots z_K$ are zeros of the system. $p_0 \cdots p_L$ are poles of the system. z_i for $i = 0 \cdots K$ can be either real or complex valued. If z_i is complex valued, then its complex conjugate should also be within zeros of the system. This is also the case for the poles p_i for $i = 0 \cdots L$. Therefore, a pole is either real valued or available with its complex conjugate.

Poles of the LTI system decide on its stability. A stable system has all its poles within the unit circle ($|z| < 1$) in complex plane. Derivation of this property can be found in Ogata (1995).

4.5.3 Stability Analysis of an LTI System in Code

As mentioned in the previous section, location of the poles play a crucial role on deciding the stability of an LTI system. Hence, if we can find poles of the system in code, then we can observe its stability. To do so, we will benefit from the available control systems libraries.

The MicroPython control systems library has two functions as `pole()` and `zero()` to extract poles and zeros of a given system. Besides, it also has the `is_stable()` function which directly tells whether the system at hand is stable or not. Let us check stability of the system in Example 4.9.

Example 4.22 *(Checking stability of an LTI system by the MicroPython control systems library)*

We can check the stability of the LTI system in Example 4.9 by the Python code in Listing 4.11 on PC. As this code is executed, we will obtain one system pole at 0.9797. Based on the location of this pole, we know that the system is stable. The function `g.is_stable()` also provides the same answer.

Listing 4.11: Checking stability of an LTI system by the MicroPython control systems library.

```
import mpcontrolPC

# System
Ts = 1
num = [0.2128]
den = [1, -0.9797]
g = mpcontrolPC.tf(num, den, Ts)

#poles of the system
zp = g.pole()

#zeros of the system
zz = g.zero()

#check stability of the system
g.is_stable()
```

The Python control systems library also has two functions as `pole(sys)` and `zero(sys)` to extract poles and zeros of a given system `sys`. We can decide on stability of the LTI system by looking at the magnitude of poles. The reader can check the usage of these functions through the Python control systems library website.

4.6 Application I: Acquiring Digital Signals from the Microcontroller, Processing Offline Data

We have two applications in this chapter. In the first application, we apply offline signal generation and processing on the STM32 microcontroller and send data to PC. The received data can be processed, such as plotting and saving to a file, by

Figure 4.18 Hardware setup for the first application.

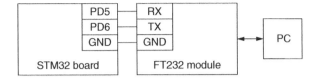

a Python code on PC. In the second application, we acquire real-time data from the STM32 microcontroller and send it to PC. As in the first application, we can process the received data by the same Python code on the PC side. Let us start with the first application.

4.6.1 Hardware Setup

We will use the FT232 module to send data to PC. To do so, the reader should connect the PD5, PD6, and any ground pin of the microcontroller to RX, TX, and ground pin of the FT232 module, respectively. Then, the FT232 module should be connected to PC using a USB cable. Finally, necessary configurations on the PC side should be done as described in Section 2.3.4. We provide the hardware setup for this application in Figure 4.18.

4.6.2 Procedure

As the reader forms the hardware setup, the C or Python code given in the next section can be executed. In the first part of this application, a parabolic input signal with 50 elements is sent to PC from the STM32 microcontroller. On the PC side, the received data is saved by the Python code and it can be plotted afterward. In the second part of this application, a step signal with 500 elements is constructed on the STM32 microcontroller. Then, this signal is applied to the LTI system given in Example 4.9. Afterward, both the input and output signals obtained from the STM32 microcontroller are sent to PC. As in the first part of the application, these data are saved and can be plotted afterward.

4.6.3 C Code for the System

For the first part of the application, we form a project in Mbed Studio and use the C code in Listing 4.12. We select the signal to be generated as parabolic.

Listing 4.12: The C code to be used in receiving an offline signal.

```
#include "mbed.h"
#include "Ccontrol.h"

Serial pc(USBTX, USBRX, 115200);
RawSerial device(PD_5, PD_6, 921600);
```

```
#define N 50

float signal[N];
char send;

int main()
{
    while(true) {

    while(device.getc() != 'r');
    parabolic_signal(N, 1, signal);
    send_data(signal,N);

    }
}
```

The C code in Listing 4.12 can be run on the STM32 microcontroller. The function `send_data()` in this code waits to send the signal entries to PC. However, this operation is not done automatically. There should be another Python code running on the PC side to trigger this operation. To do so, the reader should use the Python code in Listing 4.13.

Listing 4.13: Saving data on the PC side, part 1.

```
import mpcontrolPC
import matplotlib.pyplot as plt

N = 50
signal = mpcontrolPC.get_data_from_MC(N, 'com10', 0)
mpcontrolPC.save_to_file(signal, file_name='signal.dat')
```

The function of interest in Listing 4.13 is `get_data_from_MC(NoElements, com_port, selection)`. It is used to trigger the data transfer operation between the microcontroller and PC. Here, the parameter `NoElements` stands for the number of samples to be transferred. The parameter `com_port` is used to define which communication port is used on the PC side for data transfer. Finally, the `selection` value is used to decide on data transfer type. We should set it to 0 for offline data transfer. Then, the function `save_to_file()` is used to save the received data to a file. We can also plot the received data using available functions in the `matplotlib` library.

For the second part of the application, we form a new project in Mbed Studio and use the C code in Listing 4.14. As we execute this code, the LTI system is formed within the microcontroller. We feed a step input signal to it and observe its output.

Listing 4.14: The C code to be used in simulating a system and receiving offline signals.

```
#include "mbed.h"
#include "Ccontrol.h"
```

```
Serial pc(USBTX, USBRX, 115200);
RawSerial device(PD_5, PD_6, 921600);

#define N 500

tf_struct tf_system;
float Ts = 1;
float num[1] = {0.2128};
float den[2] = {1, -0.9797};
float input[N], output[N];

int main()
{
    //unit_pulse_signal(N, input);
    step_signal(N, 1, input);

    create_tf(&tf_system, num, den, sizeof(num)/sizeof(num[0]),
        sizeof(den)/sizeof(den[0]), Ts);
    lsim(&tf_system, N, input, output);

    while (true) {
    while(device.getc() != 'r');
    send_data(input,N);

    while(device.getc() != 'r');
    send_data(output,N);
    }
}
```

In Listing 4.14, we send both input and output signals to PC for further processing. On the PC side, we should use the Python code in Listing 4.15. Here, we
use the function get_data_from_MC(NoElements, com_port, selection) twice to receive input and output signals separately. As we receive both
signals, we can save them to separate files and plot them either on the same figure
or in different figures.

Listing 4.15: Saving data on the PC side, part 2.

```
import mpcontrolPC
import matplotlib.pyplot as plt

N = 500
signal1 = mpcontrolPC.get_data_from_MC(N, 'com10', 0)
signal2 = mpcontrolPC.get_data_from_MC(N, 'com10', 0)

mpcontrolPC.save_to_file(signal1, file_name='input.dat')
mpcontrolPC.save_to_file(signal2, file_name='output.dat')
```

4.6.4 Python Code for the System

We can repeat the first and second parts of the first application in MicroPython
on the STM32 microcontroller. For the first part of the application, the reader can
use the code in Listing 4.16. As in the C version of this application, we select the
signal to be generated as parabolic. Again, the function send_data() is used to

send data to PC. Obtaining data, plotting the result and saving data to a file steps are same as explained in Section 4.6.3.

Listing 4.16: The Python code to be used in receiving an offline signal.

```python
import pyb
import mpcontrol
import struct

uart2 = pyb.UART(2, 921600)

N = 50

def send_data(array, length):
        for i in range(length):
                dummy_input = int(struct.unpack('<I',
                    struct.pack('<f', array[i]))[0])
                dummy_send = dummy_input & 0x000000ff
                uart2.writechar(dummy_send)
                dummy_send = (dummy_input & 0x0000ff00) >> 8
                uart2.writechar(dummy_send)
                dummy_send = (dummy_input & 0x00ff0000) >> 16
                uart2.writechar(dummy_send)
                dummy_send = (dummy_input & 0xff000000) >> 24
                uart2.writechar(dummy_send)

def main():
        while True:
                while uart2.readchar() != 0x72:
                        pass
                signal = mpcontrol.parabolic_signal(N, 1)
                send_data(signal,N)
main()
```

The reader can use the Python code in Listing 4.17 for the second part of the first application. Here, we can again benefit from the setup provided in Section 4.6.3 to save and plot received data on the PC side.

Listing 4.17: The Python code to be used in simulating a system and receiving offline signals.

```python
import pyb
import mpcontrol
import struct

uart2 = pyb.UART(2, 921600)

N = 500
Ts = 1

def send_data(array, length):
        for i in range(length):
                dummy_input = int(struct.unpack('<I',
                    struct.pack('<f', array[i]))[0])
                dummy_send = dummy_input & 0x000000ff
                uart2.writechar(dummy_send)
                dummy_send = (dummy_input & 0x0000ff00) >> 8
                uart2.writechar(dummy_send)
                dummy_send = (dummy_input & 0x00ff0000) >> 16
                uart2.writechar(dummy_send)
```

```
                dummy_send = (dummy_input & 0xff000000) >> 24
                uart2.writechar(dummy_send)
def main():
        input = mpcontrol.step_signal(N, 1)
        tf_system = mpcontrol.tf([0.2128], [1, -0.9797], Ts)
        output = mpcontrol.lsim(tf_system, input, N)
        while True:
                while uart2.readchar() != 0x72:
                        pass
                send_data(input,N)
                while uart2.readchar() != 0x72:
                        pass
                send_data(output,N)
main()
```

4.6.5 Observing Outputs

We provide the obtained signal in Figure 4.19 when the first part of the application is executed. The same result can be obtained when the Python code is used instead of the C code. Therefore, we did not repeat the procedure again.

We provide the obtained signals in Figure 4.20 when the second part of the application is executed. The same result can be obtained when the Python code is used instead of the C code. Therefore, we did not repeat the procedure again.

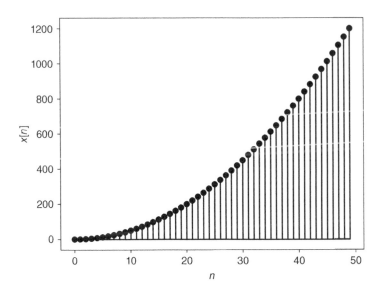

Figure 4.19 Parabolic signal obtained by the C code.

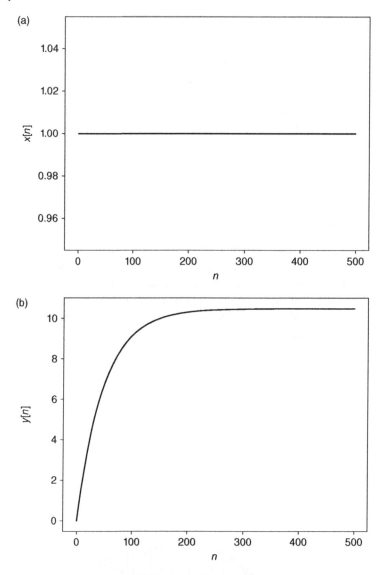

Figure 4.20 Step input signal and simulated system output obtained by the C code. (a) Input signal. (b) Output signal.

4.7 Application II: Acquiring Digital Signals from the Microcontroller, Processing Real-Time Data

We can also acquire real-time data from the microcontroller for further processing on PC. This is especially important to observe sensor data. The general system setup should be formed for this purpose as in Figure 4.21. Here, $y[n]$ is the system output, $y'[n]$ is the microcontroller input and amplifier is the gain constant between $y[n]$ and $y'[n]$.

In this application, we apply an input signal to the DC motor; obtain speed output from quadrature encoder; and send obtained input–output data to PC in real-time. The received data can be processed, such as plotting and saving to a file, by a Python code on PC.

4.7.1 Hardware Setup

For the second application, we should place the DC motor drive expansion board onto headers of the STM32 board. Then, the second upper pin of the connector CN9 of motor driver must be connected to the fourth upper pin of the connector CN9. Afterward, we should connect the positive and negative pins of the power supply module to the V_{in} and GND pins of the motor drive expansion board, respectively. The power supply should be set to 12 V. Then, we should connect the DC motor's red and black cables to the motor driver expansion board's A+ (motor terminal) and A− (other motor terminal) inputs, respectively. Afterward, DC motor's green, blue, yellow, and white cables should be connected to GND, V_{DD}, PB9, and PA0 pins of the microcontroller, respectively. We will again use the FT232 module to send data to PC with the same configuration as in the first application. We provide the hardware setup for this application in Figure 4.22.

4.7.2 Procedure

As the reader forms the hardware setup given in the previous section, the C or Python code to be given in the next section can be executed. Hence, we will feed a

Figure 4.21 General system setup for real-time data processing.

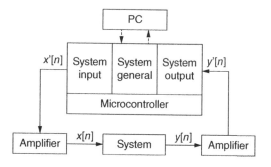

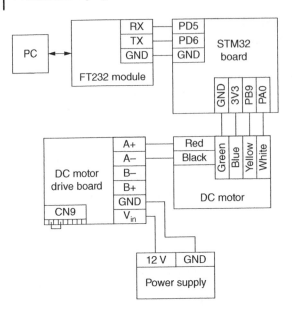

Figure 4.22 Hardware setup for the second application.

sinusoidal input signal with amplitude 3 V, frequency 5 Hz, phase 0 rad, and offset 7 V to the DC motor. Afterward, speed of the motor will be read by the quadrature encoder. The 500 elements of the sinusoidal input signal and obtained output signal will be sent to PC from the STM32 microcontroller in real-time sample by sample. On the PC side, the received data will be saved by the Python code. It can be plotted afterward if desired.

4.7.3   C Code for the System

Here, we form a new application to acquire speed of the DC motor for the selected input signal. We also let the user select the input signal and observe how the DC motor responds to it. The software setup here will be the same as in Section 3.4.3.

Within the file `system_general.cpp`, the functions `void send_data_RT(int select)` and `void send_variable(float var)` are used to send data to PC as we mentioned in Chapter 3. Here, the function `send_variable(float var)` is used to send data to PC at every sampling period. If `select` input of the function `void send_data_RT(int select)` is 0, then only input signal is sent to PC. If `select` is 1, then only the output signal is sent to PC. If `select` is 2, then both signals are sent to PC. The function `send_data_RT` is used to start data transmission process according to the `select` parameter.

As we check the `system_output.cpp` file, there are nine functions. These are briefly explained as follows: `isr_signal1rise`, `isr_signal1fall`, isr_signal2rise, and isr_signal2fall are interrupt callback functions for the quadrature encoder signals. Here, direction of the motor and the time difference between rising edge of the encoder signals are obtained. system_output_init(int type, float gain) function is used to initialize the system output block according to the type and gain inputs. type is used to select desired module to obtain system output. The user can select between ADC and quadrature encoder interface by choosing the type as 0 or 1, respectively. gain is used to set the constant gain value between system output voltage and microcontroller input voltage if ADC is selected. If the quadrature encoder interface is selected, gain value represents the constant encoder count value obtained from the motor in one revolution. obtain_system_output(void) function is used to obtain microcontroller input according to the selected module chosen by the type and gain values. calculate_frequency(int cycle_difference) function is used to obtain frequency of the quadrature encoder signals. This is done by the time difference obtained from the isr_signal1rise function. calculate_speed(float freq) is used to obtain motor speed in rpm using quadrature signal frequency, estimated motor direction, and the constant encoder count value obtained from the motor in one revolution. get_ADC function is related to ADC to obtain desired system output. This function will not be used in this application.

We should add the function send_data_RT(2), available under system_general.cpp, to the main function of our application to allow sending both motor input voltage and motor output speed to PC. Also system_output_init(1, 3591.84) function is added to initialize system output block to obtain motor output speed using quadrature encoder interface. It can be seen that the constant encoder count value obtained from our motor in one revolution is 3591.84. The modified file for our application is as in Listing 4.18.

Listing 4.18: The C code to be used in running the DC motor application.

```
#include "mbed.h"
#include "system_general.h"
#include "system_input.h"
#include "system_output.h"

Serial pc(USBTX, USBRX, 115200);
RawSerial device(PD_5, PD_6, 921600);
Ticker sampling_timer;
InterruptIn button(BUTTON1);
Timer debounce;
DigitalOut led(LED1);
PwmOut pwm1(PE_9);
PwmOut pwm2(PE_11);
DigitalOut pwm_enable(PF_15);
InterruptIn signal1(PA_0);
InterruptIn signal2(PB_9);
Timer encoder_timer;
AnalogIn ADCin(PA_3);
```

```
AnalogOut DACout(PA_4);

int main()
{
    user_button_init();
    system_input_init(1, 12/3.3);
    system_output_init(1, 3591.84);
    send_data_RT(2);
    sampling_timer_init();
    menu_init();
    while (true) {
        system_loop();
    }
}
```

As the C code in Listing 4.18 is compiled and embedded on the STM32 micro-controller, it can be executed. Hence, various input signals can be applied to the DC motor through the program interface on the Mbed Studio terminal. There is one important issue here. We can only feed a voltage value between 0 and 12 V. Therefore, the input signal should only take values between this range. Otherwise, the voltage will be saturated by the limit values.

After the C code in Listing 4.18 is executed and sinusoidal signal is selected as input with desired parameters, the system waits for the user button to start the DC motor. In order to send data to PC in real-time, the reader should run the Python code in Listing 4.19. Then, the user button should be pressed. Afterward, the DC motor starts running and data transfer to PC begins.

Listing 4.19: The Python code to be used in receiving a signal.

```
import mpcontrolPC
import matplotlib.pyplot as plt

N = 500
signal1, signal2 = mpcontrolPC.get_data_from_MC(N, 'com10', 1)
mpcontrolPC.save_to_file(signal1, file_name='input.dat')
mpcontrolPC.save_to_file(signal2, file_name='output.dat')
```

The function of interest in Listing 4.19 is `get_data_from_MC(NoElements, com_port, selection)`. Here, we should set it for real-time data transfer for two signals. Then, the function `save_to_file()` is used to store received data to desired file. We can also plot the received data.

4.7.4 Python Code for the System

We can repeat the application by embedding the Python code on the STM32 microcontroller. To do so, we should replace the content of the `main.py` file by the Python code in Listing 4.20 by following the steps explained in Section 3.2.2. We should also add the `system_general.py`, `system_input.py`, `system_output.py`, and `mpcontrol.py` files to flash. To note here, these files are provided in the accompanying book website.

Listing 4.20: The Python code to be used in running the DC motor application.

```python
import system_general
import system_input
import system_output

def main():
        system_general.button_init()
        system_general.PWM_frequency_select(0)
        system_general.sampling_timer_init(0)
        system_general.input_select(input_type='Sinusoidal Signal',
            amplitude=3, frequency=5, phase=0, offset=7, sine_select
            =0)
        system_general.motor_direction_select(1)
        system_input.system_input_init(2, 3.63636363636)
        system_output.system_output_init(1, 3591.84)
        system_general.send_data_RT(2)
        while True:
                system_general.system_loop()
main()
```

The file `system_general.py` has the functions `send_data_RT(select)` and `send_input_output(input, output, send_select)` related to sending data to PC as we mentioned in Chapter 3. Here, the function `send_input_output(input, output, send_select)` is used to send data to PC at every sampling period. If `send_select` is 0, then only input signal is sent to PC. If `send_select` is 1, then only the output signal is sent to PC. If `send_select` is 2, then both signals are sent to PC. The function `send_data_RT` is used to start data transfer according to the select parameter. If we set the select parameter as 0, then only input signal is sent to PC. If the parameter is set as 1, then only the output signal is sent to PC. If the parameter is set as 2, then both signals are sent to PC.

As we check at the `system_output.py` file, there are six functions. These are briefly explained as follows: `ic_cb(tim)` is the callback function for quadrature encoder signals. Here, direction of the motor and the time difference between rising edge of the encoder signals are obtained. `system_output_init(type, gain)` function is used to initialize the system output block according to `type` and `gain` inputs. `type` is used to select desired module to obtain system output. The user can select between ADC and quadrature encoder interface by choosing the `type` as 0 or 1, respectively. `gain` is used to set the constant gain value between system output voltage and microcontroller input voltage if ADC is selected. If quadrature encoder interface is selected, `gain` value represents the constant encoder count value obtained from the motor in one revolution. `obtain_system_output(type, gain)` function is used to obtain microcontroller input according to selected module chosen by `type` and `gain` values. `calculate_frequency(cycle_difference)` function is used to obtain frequency of the quadrature encoder signals. This is done by the time difference obtained from the function `ic_cb.calculate_speed(freq,` motor_dir_est, pulse_per_revolution) is used to obtain motor speed in rpm using quadrature signal frequency, estimated motor direction and constant encoder count value obtained from the motor in one revolution. get_ADC function is related to ADC to obtain the desired system output. This function will not be used in this application.

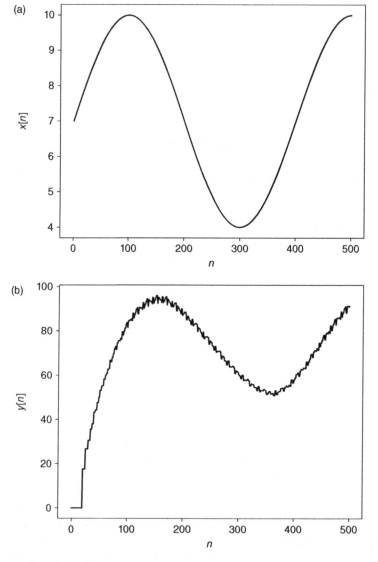

Figure 4.23 Input signal and actual DC motor speed signal obtained by C code. (a) Input signal. (b) Output signal.

We should add the function `send_data_RT(2)`, which is available under `system_general.py`, to the `main` file to allow sending both motor input voltage and motor output speed to PC. Also, the function `system_output_init(1, 3591.84)` should be added to initialize the system output block in order to obtain motor output speed using the quadrature encoder interface. It can be seen that the constant encoder count value obtained from our motor in one revolution is 3591.84. The operations on the PC side should be the same as in Section 4.7.3.

4.7.5 Observing Outputs

Figure 4.23 shows the input signal and actual output signal obtained from the DC motor. Here, we used the C code implementation in obtaining the actual input and output signals. We will use such signals in the following chapters to assess the digital controller performance. We can obtain similar results when we execute the Python code for this application as well.

4.8 Summary

Digital signals and systems form the basis of this book. Therefore, understanding them is extremely important. We introduced digital signals and systems in this chapter. Then, we evaluated digital system properties. We focused on LTI systems since they will be extensively used in this book. We also reviewed the z- and inverse z-transform in this chapter. We provided Python based z- and inverse z-transform calculation methods on PC. We also analyzed the properties of digital LTI systems via z-transform. As end of chapter applications, we provided two cases of data acquisition from the STM32 microcontroller. The first application focuses on offline data processing. The second application provides methods of acquiring real-time data from the microcontroller. We will frequently use both applications in the following chapters.

Problems

4.1 Generate the following digital signals, with parameters of your choice, for their first 20 elements in Python on PC. Plot them using the `matplotlib` library.
   a. Pulse.
   b. Step.
   c. Ramp.
   d. Parabolic.
   e. Exponential.

    f. Sinusoidal.

    g. Damped sinusoidal.

    h. Rectangular

    i. Sum of sinusoids.

    j. Sweep.

    k. Random.

4.2 Repeat Problem 4.1 in C language on the STM32 microcontroller. Transfer the generated signals to PC via UART communication. Plot them using the `matplotlib` library in Python.

4.3 Represent the below LTI systems by the Python control systems library.

    a. $G_1(z) = \frac{1}{5z^3 - 3z^2 + 2z + 1}$

    b. $G_2(z) = \frac{0.3625z}{(z-1)(z-0.4562)}$

    c. $G_3(z) = \frac{0.006147z + 0.005953}{z^2 - 1.907z + 0.9081}$

4.4 Obtain and plot the unit step response of LTI systems in Problem 4.3 in Python.

4.5 Repeat Problem 4.3 via

    a. MicroPython control systems library.

    b. C control systems library on the STM32 microcontroller.

4.6 Repeat Problem 4.4 via

    a. MicroPython control systems library.

    b. C control systems library on the STM32 microcontroller.

4.7 Obtain the constant-coefficient difference equation of the LTI systems with transfer function in Problem 4.3.

4.8 Use the MicroPython and Python control systems libraries separately, to connect LTI systems in Problem 4.3 with below setups.

    a. $G_1(z)$ is serially connected to $G_2(z)$.

    b. $G_2(z)$ and $G_3(z)$ is connected in parallel.

    c. $G_3(z)$ and $G_1(z)$ form a negative feedback loop.

4.9 Obtain the inverse z-transform of $G_1(z)$, $G_2(z)$, and $G_3(z)$ in Problem 4.3 using Python functions.

4.10 Check stability of LTI systems in Problem 4.3 using MicroPython control systems library.

5

Conversion Between Analog and Digital Forms

This book is on digital control. However, actual control signals are in analog form most of the times. These should be represented in digital form so that they can be processed by the STM32 microcontroller. As the digital signal is processed or desired control signal is generated by the microcontroller, it should be converted to analog form. Hence, it can be applied to the analog system to be controlled. For both cases, we need methods for converting analog and digital signals from one to other. There are solid mathematical theories explaining how these conversions can be done. We will start with explaining them in this chapter. Besides theory, one needs practical methods to convert analog and digital signals. Modern microcontrollers (as candidate digital control system platforms) have dedicated hardware for this purpose. We introduced such hardware on the STM32 microcontroller in Chapter 2. We will focus on how this hardware can be used both in Python and C languages in this chapter. Hence, the reader will be able to process an analog signal on the microcontroller and feed the processed signal to outside world in analog form.

The user may have an analog controller working as desired which has been designed beforehand. We may want to represent it in digital form as code. Hence, we can duplicate the existing analog controller in digital form such that no analog parts are needed for its implementation. This way, we can benefit from the advantages of digital domain. In a similar manner, most existing systems to be controlled are also in analog form. We may need their digital representations for control operations. Therefore, it becomes a necessity to bridge analog and digital domains. To do so, we will focus on methods to represent an analog system in digital form in this chapter. At the end of chapter, we will also provide a practical application emphasizing all the methods considered here.

Embedded Digital Control with Microcontrollers: Implementation with C and Python,
First Edition. Cem Ünsalan, Duygun E. Barkana, and H. Deniz Gürhan.
© 2021 The Institute of Electrical and Electronics Engineers, Inc. Published 2021 by John Wiley & Sons, Inc.
Companion website: www.wiley.com/go/Unsalan/Embedded_Digital_Control_with_Microcontrollers

5.1 Converting an Analog Signal to Digital Form

Analog to digital conversion (ADC) operation should first be understood by mathematical derivations. Then, the reader should learn how to realize the ADC operation in practice. Therefore, we start with the mathematical derivation of ADC followed by its implementation via Python and C languages in this section.

5.1.1 Mathematical Derivation of ADC

Before explaining ADC, we should inform the reader that we will be using the discrete-time signal representation introduced in Chapter 4 in mathematical derivation. As a reminder, the discrete-time signal lies between analog (continuous-time) and digital signal representations. The distinction between discrete-time and digital signals is as follows. The discrete-time signal samples may take real (unlimited) values. On the other hand, the digital signal samples can only take limited values.

Let us start with the analog signal $x(t)$. Assume that, we want to represent it in digital domain. Therefore, we will need samples of the analog signal in time. This operation is well established by the sampling theorem (Oppenheim and Schafer 2009). Although mathematical derivation of this theorem is beyond the scope of this book, we briefly explain its working principles. To do so, we should first introduce the impulse signal, $\delta(t)$, defined in continuous-time. This is a signal with amplitude going to infinity and width shrinking to zero as $t \to 0$. For all other index t, the impulse signal has value zero. The area below the impulse signal is one. This signal serves as a perfect candidate to represent the sampling operation in mathematical terms.

In order to obtain samples of the analog signal $x(t)$ in time, we can multiply it by an impulse train defined as $\delta_T(t) = \sum_{n=-\infty}^{\infty} \delta(t - nT_s)$. Before multiplication, let us first understand the properties of impulse train. This signal is composed of infinite number of impulse signals located T_s time instants apart. Hence, it is a periodic signal with period T_s. Multiplying a continuous-time signal by an impulse leads to another impulse signal (Oppenheim and Schafer 2009). This new impulse will have a multiplier having value of the multiplied signal at the time instant the impulse signal is defined. In other words, if we multiply a continuous-time signal $x(t)$ by $\delta(t - T_s)$, we will get $x(T_s)\delta(t - T_s)$. This is called sifting theorem.

We can multiply the analog signal $x(t)$ by $\delta_T(t)$ and obtain $x(nT_s) = \sum_{n=-\infty}^{\infty} x(nT_s)\delta(t - nT_s)$. As a result, samples of the continuous-time signal $x(t)$ appear as coefficient of impulse signals periodically located in time by T_s. However, the signal $x(nT_s)$ is still in continuous-time. If we just focus on the coefficient of impulse signals with their location as index, we will have $\tilde{x}[n] = x(nT_s)$. Here, $\tilde{x}[n]$ represents the discrete-time signal obtained from $x(t)$.

Now, the question arises. What should be the value of T_s such that the obtained discrete-time signal represents its analog version as best as it can? The sampling theorem gives the answer. In order to explain it, let us start with two definitions. First, period of the impulse train can also be represented as frequency $f_s = 1/T_s$. We can call this entity as sampling frequency since we are taking samples of the continuous-time signal with this frequency. Second, a continuous-time signal should have a maximum frequency value such that its magnitude content is zero beyond this range. This is called bandwidth of the signal. Let us call the bandwidth of $x(t)$ as Ω_M rad/sec. Sampling theorem tells us that if we pick $2\pi f_s > 2\Omega_M$, then we do not lose any information during sampling (Oppenheim and Schafer 2009). To note here, the sampling theorem does not tell us that an analog signal with $\Omega_M \to \infty$ cannot be sampled. It only tells us that if the mentioned conditions in the theorem are satisfied, then no information will be lost during the sampling operation.

The sampling operation only focuses on obtaining samples of the analog signal in time. As mentioned before, values of the obtained signal will still be real. In order to represent these values in a digital system, we should quantize them. The quantization operation can be described in mathematical terms as follows. Let us pick a sample as $\tilde{x}[n_0] = x(n_0 T_s)$. First, we should obtain the range (minimum and maximum values) of the signal $x(nT_s)$. Assume these values are X_{min} and X_{max}, respectively. Next, we should set the quantization level which represents how many bits will be assigned per sample. Let us call this value as N. The resulting digital value for the analog input signal sample can be calculated as

$$x[n_0] = \left\lfloor \frac{(2^N - 1)(\tilde{x}[n_0] - X_{min})}{(X_{max} - X_{min})} \right\rfloor \tag{5.1}$$

where $\lfloor \cdot \rfloor$ stands for the floor function. Please note that we will have an integer value between 0 and $2^N - 1$ after quantization. We can use this value in two different ways. First, we can convert it to float representation by normalizing it (by N and an appropriate multiplier) and perform floating-point operations as we have been doing till now. Second, we can convert the integer value to a fixed-point representation and use it accordingly (Ünsalan et al. 2018).

The difference (in terms of the analog value) between two successive quantization levels is called resolution. Naturally, higher the quantization level, better the resolution. However, as the number of bits assigned to each sample increases, memory space required to save them becomes problematic. Moreover, the CPU clock speed should be increased to process more samples in a given time slot. Therefore, balance needs to be established between the resolution and data size that can be stored and processed. For more information on these topics, please see the book by Proakis and Manolakis (1995).

Sampling and quantization operations are jointly called ADC. Next, we will consider this operation from a practical perspective both in Python and C languages.

5.1.2 ADC in Code

Although sampling theorem clearly indicates how the sampling operation should be done, it may not be possible to achieve this in practice. This is also the case for the quantization operation. The main reason for the difference between the theory and practice is that the ADC operation is highly dependent on the hardware used. Therefore, it may not be possible to set the desired sampling frequency due to hardware limitations. Moreover, available memory and CPU speed may limit the number of samples that can be processed by the microcontroller in a given time slot. As a final limitation, sampling theorem assumes the analog signal to be sampled is bandlimited. This is rarely the case for actual signals in real life. Therefore, theoretical requirements may not be satisfied in practice. However, we can still apply the ADC operation and obtain a reasonable digital signal corresponding to its analog version.

We handled the ADC operation on the STM32 microcontroller using MicroPython in Section 3.2.3. We guide the reader to that section on practical ADC operations to avoid duplication. Here, we provide a sample code to modify the sampling rate and quantization level in Python language in Listing 5.1. Input signal for this code is provided by an external signal generator. To do so, the positive connector of the signal generator should be connected to PA6 pin of the STM32 board. The ground connector of the signal generator should be connected to any ground pin of the board. Then, the reader should generate a sinusoidal signal with frequency 1 Hz and range between 0.5 and 2.5 V.

Listing 5.1: Changing the sampling period and quantization level for the ADC operation in MicroPython.

```
import pyb
from array import array
import struct

N = 500

uart2 = pyb.UART(2, 921600)
Timer1 = pyb.Timer(1, freq = 100, mode = pyb.Timer.UP)
adc_A6 = pyb.ADC(pyb.Pin('PA6'))
LD2 = pyb.Pin('PB7', mode=pyb.Pin.OUT_PP)

interrupt_cnt = 0
print_flag = 0
buffer = array('H', (0 for i in range(N)))
#buffer = array('B', (0 for i in range(N)))
buffer_f = array('f', (0 for i in range(N)))

def send_data(array, length):
```

```
            for i in range(length):
                dummy_input = int(struct.unpack('<I', struct.pack
                        ('('<f', array[i]))[0])
                dummy_send = dummy_input & 0x000000ff
                uart2.writechar(dummy_send)
                dummy_send = (dummy_input & 0x0000ff00) >> 8
                uart2.writechar(dummy_send)
                dummy_send = (dummy_input & 0x00ff0000) >> 16
                uart2.writechar(dummy_send)
                dummy_send = (dummy_input & 0xff000000) >> 24
                uart2.writechar(dummy_send)
def convert_ADCvalue_to_voltage(value):
        voltage = 3.3 * value / 4095
#       voltage = 3.3 * value / 255
        return voltage

def get_ADC(timer):
        global print_flag
        print_flag = 1


def main():
        global interrupt_cnt
        global print_flag

        Timer1.callback(get_ADC)
        while True:
                if print_flag == 1:
                        if interrupt_cnt ('< N:
                                buffer[interrupt_cnt] = adc_A6.
                                        read()
#                               buffer[interrupt_cnt] = adc_A6.
                                        read() >> 4
                                interrupt_cnt = interrupt_cnt + 1
                        else:
                                LD2.value(1)
                                Timer1.deinit()
                                for i in range(N):
                                        buffer_f[i] =
                                                convert_ADCvalue_to_
                                                voltage(buffer[i])
                                while uart2.readchar() != 0x72:
                                        pass
                                send_data(buffer_f,N)
                        print_flag = 0
main()
```

In Listing 5.1, we use the Timer1 module to trigger the ADC module. Here, we select the trigger period (sampling period) as 0.01 seconds. The reader can change this value for his or her needs to set the sampling period. The acquired data this way is stored in a buffer array, with 500 elements, for five seconds. As the array is filled, the timer stops. The values are converted to voltage levels between 0 and 3.3 V. The LD2 LED turns on to indicate that the operation ended. Then, the reader can run the below code snippet on PC to transfer the acquired data to PC. Then, this data can be plotted if required.

```
import mpcontrol
import mpcontrolPC
import matplotlib.pyplot as plt
N = 500
signal = mpcontrolPC.get_data_from_MC(N, 'com10', 0)
mpcontrolPC.save_to_file(signal, file_name='ADC.dat')
```

The quantization level in the STM32 microcontroller is fixed to 12 bits. However, we can postprocess the acquired data and decrease the quantization level. We provide sample usage of this procedure in Listing 5.1. Here, the reader can replace the commented out sections of the code to obtain 8-bit data.

The reader can reach the necessary information on practical ADC operations in C language in Section 3.3.3. As in MicroPython, we provide the sample C code in Listing 5.2 to modify the sampling period and quantization levels. Here, the sampling period is changed based on the ticker function. Besides, all operations here are the same as in the MicroPython case. As for quantization, the ADC function in Mbed provides results in 16-bit format. The reader can decrease this value to 8 bits by replacing the commented out parts in the C code.

Listing 5.2: Changing the sampling period and quantization levels for the ADC operation in C.

```
#include "mbed.h"
#include "Ccontrol.h"

#define N 500

RawSerial device(PD_5, PD_6, 921600);
Ticker timer_ticker;
AnalogIn analog(PA_6);
DigitalOut led2(LED2);

int interrupt_cnt = 0;
int print_flag = 0;
uint16_t buffer[N];
//uint8_t buffer[N];
float buffer_f[N];

float convert_ADCvalue_to_voltage(int value){
        float voltage;
        voltage = 3.3 * ((float)value) / 65535;
//      voltage = 3.3 * ((float)value) / 255;
        return voltage;
}

void get_ADC(){
    print_flag = 1;
}

int main()
{
        int i;

        timer_ticker.attach(&get_ADC, 0.01);
```

```
while(true){
        if(print_flag == 1){
    if(interrupt_cnt<N){
            buffer[interrupt_cnt] = analog.read_u16();
//          buffer[interrupt_cnt] = analog.read_u16() >> 8;
            interrupt_cnt++;
    }
    else{
        led2 = 1;
        timer_ticker.detach();
        for(i=0;i<N;i++){
            buffer_f[i] = convert_ADCvalue_to_voltage
                (buffer[i]);
        }
        while(device.getc() != 'r');
        send_data(buffer_f,N);
    }
    print_flag = 0;
        }
    }
}
```

5.2 Converting a Digital Signal to Analog Form

As the name implies, digital to analog conversion (DAC) deals with converting a digital signal to analog form. In this section, we will derive properties of DAC from a mathematical perspective first. Then, we will focus on the practical DAC implementation via Python and C languages.

5.2.1 Mathematical Derivation of DAC

While explaining DAC, we will use the discrete-time representation instead of digital form in mathematical derivations as in the previous section. Assume that we have an analog signal $x(t)$ with bandwidth Ω_M rad/s. Assume further that we have the sampled version of the analog signal as $x(nT_s)$ as in the previous section. We can pass $x(nT_s)$ through an ideal low-pass filter, with impulse response $h_{ILPF}(t)$, to reconstruct it from its samples as

$$x_r(t) = x(nT_s) * h_{ILPF}(t) \tag{5.2}$$

$$= \sum_{n=-\infty}^{\infty} x(nT_s)\delta(t - nT_s) * \frac{w_c T_s}{\pi}\text{sinc}(w_c t)$$

$$= \frac{w_c T_s}{\pi} \sum_{n=-\infty}^{\infty} x(nT_s)\text{sinc}(w_c(t - nT_s))$$

Eq. (5.2) indicates that, every sample is represented by a weighted sinc function in reconstruction (Ünsalan et al. 2018). If T_s is picked to satisfy the sampling theorem, then we will have perfect reconstruction, $x_r(t) = x(t)$, using the sum of sinc functions. This operation is called DAC. Next, we will consider it from a practical perspective both in Python and C languages.

5.2.2 DAC in Code

The mathematical derivation in the previous section tells us that perfect reconstruction is possible after sampling an analog signal. Unfortunately, this is not the case in practice. Let us explain why. First, we have the ADC module to perform the sampling operation. However, this module also quantizes amplitude of each sample besides sampling it in time domain. The quantization operation is nonreversible. Second, a continuous-time signal cannot be bandlimited in real-life. Therefore, it should be filtered by an anti-aliasing filter before the sampling operation. This filter discards high frequency components of the original signal. These distortions occur before the reconstruction operation. Third, we know that the ideal low-pass filter used to reconstruct the sampled signal cannot be realized (Ünsalan et al. 2018). Instead, a non-ideal low-pass filter should be used (such as a zero order hold circuit) in practice. This filter also distorts frequency content of the signal. Therefore, there can never be a perfect reconstruction in practice. However, we can still apply the DAC operation to a digital signal to obtain its analog version with acceptable quality.

We handled the DAC operation on the STM32 microcontroller using MicroPython in Section 3.2.3. We guide the reader to that section on practical DAC operations to avoid duplication. We can change the sampling period and quantization levels for the DAC operation following the same steps as in the ADC case. We provide such a Python code in Listing 5.3. Here, we generate a sinusoidal signal having frequency of 5 Hz and voltage levels between 0.5 and 2.5 V. The sampling period for this operation is selected as 0.001 seconds. The reader can modify the sampling period within this code for the DAC operation. The quantization level in the DAC operation is 12-bits by default. It can be changed to 8 bits by replacing the commented out code segments in Listing 5.3.

Listing 5.3: Changing the sampling period and quantization level for the DAC operation in MicroPython.

```
import pyb
from array import array
from math import pi,sin

sine_freq = 5
N = 200

Timer1 = pyb.Timer(1, freq = sine_freq*N, mode = pyb.Timer.UP)
dac_flag = 0

def convert_DACvoltage_to_value(voltage):
```

```
#          value = int(4095 * voltage / 3.3)
#          value = int(255 * voltage / 3.3)
           return value

def generate_DAC(timer):
           global dac_flag
           dac_flag = 1

def main():
           global dac_flag
           interrupt_cnt = 0
           dac1 = pyb.DAC(1, bits=12, buffering=True)
#          dac1 = pyb.DAC(1, bits=8, buffering=True)
           sine_voltages = array('f', 1.5 + sin(2 * pi * i / N) for i
                  in range(N))
           buffer = array('H', (0 for i in range(N)))
#          buffer = array('B', (0 for i in range(N)))
           for i in range(N):
                   buffer[i] = convert_DACvoltage_to_value
                          (sine_voltages[i])
           Timer1.callback(generate_DAC)
           while True:
                   if dac_flag == 1:
                          dac1.write(buffer[interrupt_cnt])
                          interrupt_cnt = interrupt_cnt + 1
                          if interrupt_cnt == N:
                                  interrupt_cnt = 0
                          dac_flag = 0
main()
```

The reader can reach the necessary information on practical DAC operations in C language in Section 3.3.3. As in the MicroPython case, we provide a C code for changing the sampling period in Listing 5.4. Unfortunately, the quantization level is fixed to 12-bits and cannot be modified. Therefore, it is left unchanged in the code.

Listing 5.4: Changing the sampling period for the DAC operation in C.

```
#include "mbed.h"

#define PI 3.14159265358979323846
#define N 200
#define sine_freq 5

Ticker timer_ticker;
AnalogOut dac(PA_4);

float sine_voltages[N];
int dac_flag = 0;

void set_DAC(){
    dac_flag = 1;
}

int main()
{
    int i;
    float period = 1 / (float)N / (float)sine_freq;
        timer_ticker.attach(&set_DAC, period);
        for(i=0;i<N;i++){
```

```
            sine_voltages[i] = 1.5 + sin(2 * PI * i / N);
    }
    i = 0;

    while(true){
            if(dac_flag == 1){
                    dac.write(sine_voltages[i]/3.3);
                    i++;
                    if(i==N)i=0;
                    dac_flag = 0;
            }
    }
}
```

5.3 Representing an Analog System in Digital Form

As we convert an analog signal to digital form (and a digital signal to analog form as well), the next step is representing an analog system in digital form. The aim here is as follows. While replacing the analog system with its digital counterpart, the end user should not observe any differences between both systems. If we can achieve this, then we can directly replace the analog system with its digital counterpart (as a code snippet running on a digital system such as microcontroller). Besides, we may want to obtain the digital form of an existing analog system to design a digital controller for it. For both cases, we need mathematical methods.

There are three approaches to obtain the digital form of an analog system. In the first approach, poles and zeros of the analog system are mapped to digital form by a transformation. Therefore, this method is called pole-zero matching. In the second approach, response of the analog system to a given step input is approximated by its sampled version. The digital system obtained using this method is called zero-order hold equivalent. In the third approach, differential equation representing the analog system is approximated by the corresponding difference equation. Hence, transfer function of the analog system can be represented in digital form. Here, the aim is making analog and digital transfer functions to be as similar as possible. We will use the bilinear transformation (also called Tustin's approximation) method for this purpose.

We will use the simple RC filter as the analog system in the following sections. Layout of the system is as in Figure 5.1. Let us take the input for this system as the input voltage represented by $x(t)$. Furthermore, let us take the output as voltage on the capacitor represented as $y(t)$. We will obtain the transfer function of this system in Chapter 6. For our applications in this chapter, we can take the analog transfer function of this system as

$$G(s) = \frac{1000}{s + 1000} \tag{5.3}$$

We will convert this transfer function to digital form in the following sections.

Figure 5.1 Simple RC filter as the analog system.

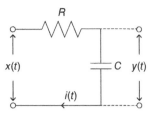

5.3.1 Pole-Zero Matching Method

To apply the pole-zero matching method, we should first obtain all poles and zeros of the continuous-time system. Then, we map these to digital domain with the transformation $z = e^{sT_s}$ where T_s is the sampling period. As a result, we reconstruct the equivalent digital representation of the system based on these mapped poles and zeros. Gain of the continuous and digital systems should also be matched. Franklin et al. (2006) suggest using $G(s)|_{s=0} = G(z)|_{z=1}$ for this purpose. For more information on this method, please check the mentioned reference.

We can benefit from the Python control systems library to obtain the digital representation of an analog system using the pole-zero matching method. To do so, we should use the function `sample_system(sysc, Ts, method,alpha)` within this library. Here, `sysc` is the representation of the analog system; `Ts` is the sampling period; `alpha` is the generalized transformation weighting parameter which is ignored for our case. `method` should be selected as `'matched'` to use the pole-zero matching method. To note here, this function does not provide the gain adjustment for the pole-zero matching method. The user should perform this operation by hand. Let us provide an example on the usage of this function.

Example 5.1 *(Representing an analog system in digital form using pole-zero matching method)*

Let us take the analog transfer function of the simple RC filter given in Eq. (5.3). We can take $T_s = 0.0005$ seconds. We can use the Python control systems library to obtain the digital representation of this system using the pole-zero matching method with the Python code in Listing 5.5.

Listing 5.5: Representing an analog system in digital form using pole-zero matching method.

```
import control

Ts=0.0005
num=[1000]
den=[1, 1000]

gs = control.TransferFunction(num, den)
gz = control.sample_system(gs, Ts, method='matched')

print('Analog system transfer function',gs)
print('Digital system transfer function',gz)
```

As we execute the code in Listing 5.5, we obtain the digital version of the simple RC filter as

$$G(z) = \frac{393.5}{z - 0.6065} \tag{5.4}$$

In Eq. (5.4), the gain of the system is not adjusted. To do so, we should use $G(s)|_{s=0} = G(z)|_{z=1}$. Hence, we should scale Eq. (5.4) by $393.5/0.3935$ and obtain

$$G(z) = \frac{0.3935}{z - 0.6065} \tag{5.5}$$

5.3.2 Zero-Order Hold Equivalent

Within the zero-order hold equivalent method, both analog and digital systems will have the same output to unit step input. This is important since digital controllers can be analyzed based on their step response as will be seen in Chapter 7.

Let us explain how the zero-order hold equivalent method works. Assume that we have an analog system with impulse response $g(t)$ and digital system with impulse response $g[n]$. The sampling period is T_s. We want both systems to give the same response for the unit step input signal at sampling instants. In other words, we want $(g(t) * u(t))_{t=nT_s} = g[n] * u[n]$. We can represent this equality in complex domain using Laplace and z-transforms as

$$G(z)\frac{1}{1 - z^{-1}} = \mathcal{Z}\left(\mathcal{L}^{-1}\left(\frac{G(s)}{s}\right)_{t=nT_s}\right) \tag{5.6}$$

where $\mathcal{Z}$ and $\mathcal{L}^{-1}$ represent the z-transform and inverse Laplace transform, respectively. We can leave $G(z)$ alone in Eq. (5.6) and obtain

$$G(z) = (1 - z^{-1})\mathcal{Z}\left(\mathcal{L}^{-1}\left(\frac{G(s)}{s}\right)_{t=nT_s}\right) \tag{5.7}$$

As a result, we can obtain the digital representation, $G(z)$, of the analog system $G(s)$.

We can benefit from the Python control systems library to obtain the digital representation of an analog system using the zero-order hold equivalent method. To do so, we should use the function `sample_system(sysc, Ts, method,alpha)` within this library as in the previous section. Here, the method should be selected as `'zoh'` to use the zero-order hold equivalent. Let us provide an example on the usage of this function.

Example 5.2 (Representing an analog system in digital form using zero-order hold equivalent method)
We can repeat the process in Example 5.1 for the zero-order hold equivalent method. To do so, we should use the Python code in Listing 5.5 by modifying it

as `sample_system(gs, Ts, method='zoh')`. As we execute the code, we obtain digital version of the simple RC filter as

$$G(z) = \frac{0.3935}{z - 0.6065} \tag{5.8}$$

Here, the obtained transfer function is the same as in Eq. (5.5) obtained by the pole-zero matching method.

5.3.3 Bilinear Transformation

Bilinear transformation (or Tustin's approximation) method is based on approximating a definite integral by the sum of trapezoids. Therefore, differential equation representing the analog system should first be converted to the corresponding integral equation. Afterward, integral approximation by sum of trapezoids can be performed. This method is applied on a simple differential equation in Ogata (1995). As a result, the required conversion is obtained as

$$s = \frac{2}{T_s} \frac{1 - z^{-1}}{1 + z^{-1}} \tag{5.9}$$

As indicated by Ogata, this conversion works for all LTI continuous-time systems (Ogata 1995). Therefore, Eq. (5.9) can be used to find $G(z)$ from $G(s)$ using the bilinear transformation method.

We can benefit from the MicroPython control systems library to obtain the digital version of an analog signal. To do so, we should use the function `c2d(tfc, ts, 'bilinear')` within the library. Here, `tfc` stands for the continuous-time transfer function to be converted. `ts` is the sampling period to be used in operation. Let us consider an example on the usage of this function.

Example 5.3 *(Representing an analog system in digital form using bilinear transformation by MicroPython control systems library)*
Let us take the analog transfer function of the simple RC filter given in Eq. (5.3). We can take $T_s = 0.0005$ seconds and use the MicroPython control systems library to obtain the digital representation of the system using the bilinear transformation method with the Python code in Listing 5.6.

Listing 5.6: Representing an analog system in digital form using bilinear transformation method.

```
import mpcontrolPC

Ts=0.0005
num=[1000]
den=[1, 1000]

gs = mpcontrolPC.tf(num, den)
gz = mpcontrolPC.c2d(gs, Ts, method='bilinear')

print('Analog system transfer function',gs)
print('Digital system transfer function',gz)
```

As we execute the code in Listing 5.6, we will obtain digital version of the simple RC filter as

$$G(z) = \frac{0.2z + 0.2}{z - 0.6} \qquad (5.10)$$

As can be seen in Eq. (5.10), the obtained digital transfer function by the bilinear transformation method is different from the previous two methods introduced earlier.

We can also benefit from the Python control systems library to obtain the digital representation of an analog system using the bilinear transformation method. To do so, we should use the function `sample_system(sysc, Ts, method, alpha)` within this library. Here, `sysc` is the representation of the analog system, `Ts` is the sampling period, `method` should be selected as `'tustin'` to use the bilinear transformation in obtaining the digital form. Let us provide an example on the usage of this function.

Example 5.4 *(Representing an analog system in digital form using the bilinear transformation method)*
We can repeat the process in Example 5.3 using Python control systems library. To do so, we should use the Python code in Listing 5.5 by modifying it as `sample_system(gs, Ts, method='tustin')`. As we execute the code, we obtain digital version of the simple RC filter as

$$G(z) = \frac{0.2z + 0.2}{z - 0.6} \qquad (5.11)$$

As we compare Eqs. (5.10) and (5.11), we can observe that MicroPython and Python control systems libraries provide the same result.

5.4 Application: Exciting and Simulating the RC Filter

The aim of this application is understanding the ADC and DAC concepts in practice. In the first part of the application, an actual RC filter will be constructed. Unit step input will be fed to it from the STM32 microcontroller through its DAC module. Then, analog output of the system will be sampled by the ADC module and data will be stored. In the second part of the application, the reader will convert the analog system (RC filter) to digital form and obtain its response to unit step input as simulation. Hence, the reader will be able to compare the simulation and actual implementation results on the STM32 microcontroller. The system setup for this application will be the same as in Figure 4.21.

Figure 5.2 Hardware setup for the application.

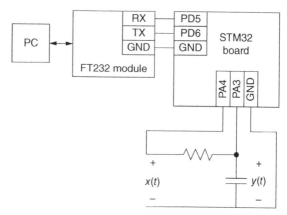

5.4.1 Hardware Setup

For this application, we can construct an actual RC filter by picking a resistor and capacitor with $R = 10$ kΩ and $C = 100$ ηF. We should connect DAC output of the STM32 microcontroller to $x(t)$ input of the RC filter through pin PA4 of the STM32 board. We should also connect $y(t)$ output of the RC filter to ADC input of the STM32 microcontroller through pin PA3 of the STM32 board. Here, we will use the FT232 module to send data to PC with the same setup described in Section 4.7. Afterward, the system will be ready to be used in the application. We provide hardware setup for the application in Figure 5.2.

5.4.2 Procedure

In the first part of the application, we will actually construct the RC filter and feed unit step signal to it from the STM32 microcontroller. To do so, we will generate the signal in the microcontroller. Then convert it to analog form by DAC module of the STM32 microcontroller. We will sample output of the RC filter by ADC module of the STM32 microcontroller and feed actual step response of the system to PC through the Python interface. Here, the sampling period should be set as $T_s = 0.0005$ seconds.

In the second part of the application, the reader should use digital form of the RC filter obtained by bilinear transformation as given in Eq. (5.10). We should feed a unit step input to this digital system and obtain the corresponding output as simulation.

5.4.3 C Code for the System

In order to form the C code for the application, we should first create a new project under Mbed Studio. As the project is generated, we should replace the content of the `main.cpp` file with the C code given in Listing 5.7. The software setup here will be the same as in Section 3.4.3.

Listing 5.7: The C code to be used in exciting the RC filter.

```c
#include "mbed.h"
#include "system_general.h"
#include "system_input.h"
#include "system_output.h"

Serial pc(USBTX, USBRX, 115200);
RawSerial device(PD_5, PD_6, 921600);
Ticker sampling_timer;
InterruptIn button(BUTTON1);
Timer debounce;
DigitalOut led(LED1);
PwmOut pwm1(PE_9);
PwmOut pwm2(PE_11);
DigitalOut pwm_enable(PF_15);
InterruptIn signal1(PA_0);
InterruptIn signal2(PB_9);
Timer encoder_timer;
AnalogIn ADCin(PA_3);
AnalogOut DACout(PA_4);

int main()
{

    user_button_init();
    system_input_init(0, 1);
    system_output_init(0, 1);
    send_data_RT(2);
    sampling_timer_init();
    menu_init();
    while (true) {
        system_loop();
    }

}
```

In Listing 5.7, we set the DAC output as system (RC filter) input via the function `system_input_init(type, gain)`. Here, `type` is set to 0 for using DAC output and `gain` is set to 1. We apply the generated voltage directly to the system. We use the function `system_output_init(type, gain)` to measure the system output by the ADC module of the STM32 microcontroller. Here, `type` is set to 0 to use the ADC input and `gain` is set to 1. We then read the system output directly. In this application, the menu on the Mbed Studio will not ask for the PWM frequency value or direction since we do not use the DC motor as a system. We should set the sampling frequency to 2 kHz ($T_s = 0.0005$ seconds). Finally, the reader should select the input signal as step, with amplitude 1 V.

After the C code in Listing 5.7 is executed and step input is selected as input with 1 V amplitude, the system waits for user button press to start. In order to send data to PC in real-time, run the Python code in Listing 5.8. Then, press the user button. Afterward, the system starts working and data transfer to PC begins.

Listing 5.8: The Python code to send data to PC in real-time.

```
import mpcontrolPC
import mpcontrol

N = 100
signal1, signal2 = mpcontrolPC.get_data_from_MC(N, 'com10', 1)
mpcontrolPC.save_to_file(signal1, file_name='input.dat')
mpcontrolPC.save_to_file(signal2, file_name='output.dat')

Ts = 0.0005
num = [1000]
denum = [1, 1000]

tf_system = mpcontrolPC.tf(num, denum)
tf_system_discrete = mpcontrolPC.c2d(tf_system, Ts)

y = mpcontrolPC.step(tf_system_discrete, stepgain=1, nsample=100)
mpcontrolPC.save_to_file(y, file_name='output_simulation')
```

The function of interest in Listing 5.8 is `get_data_from_MC(NoElements,` `com_port, selection)`. Here, we should set the `selection` to 1 for real-time data transfer for two signals. Then, continuous-time transfer function of the system is converted the digital form using bilinear transformation. Same step input is applied to this digital transfer function and simulation output is saved to the variable `y`. The function `save_to_file()` is used to store the received and simulated data to desired files. We can also plot obtained and simulated data to observe them.

5.4.4   Python Code for the System

In order to form the Python code for the application, first we should replace the content of the `main.py` file with the Python code given in Listing 5.9 by following the steps explained in Section 3.2.2. We should also place the `system_general.py`, `system_input.py`, `system_output.py`, and `mpcontrol.py` files to flash. To note here, these files are provided in the accompanying book web site.

Listing 5.9: The Python code to be used in exciting the RC filter.

```
import system_general
import system_input
import system_output

def main():
        system_general.sampling_timer_init(0)
        system_general.input_select(input_type='Step Signal',
            amplitude=1)
        system_input.system_input_init(0, 1)
        system_output.system_output_init(0, 1)
        system_general.send_data_RT(2)
        while True:
                system_general.system_loop()
main()
```

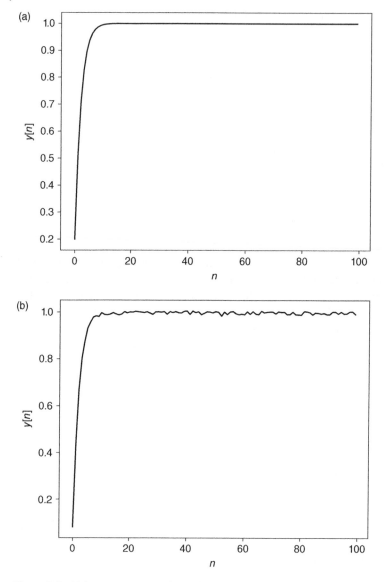

Figure 5.3 Unit step response of the RC filter, simulation and actual measurements obtained by C code. (a) Simulated output. (b) Actual output.

In Listing 5.9, we apply the same steps as in the C code given in Listing 5.7. The Python code to be used on the PC side is the same as in Section 5.4.3. As a reminder, the reader should set `Ts=0.0005` for the application.

5.4.5 Observing Outputs

As we apply the procedure explained in the previous section, we will obtain the results as in Figure 5.3. Here, response of the RC filter to unit step input is given both in terms of simulation and actual measurements obtained by C code. As can be seen in the figure, obtained output signals are the same. The Python code provided in the previous section should also give similar results.

5.5 Summary

Digital processing of continuous-time signals opens up a way to replace any analog system with its digital counterpart. This leads to digitalization of existing analog systems. The first step here is obtaining digital version of the analog signal to be processed. The next step is obtaining analog version of the processed digital signal. In this chapter, we considered these steps through ADC and DAC modules. The remaining part to be considered is representing the continuous-time system in digital form. We also focused on this issue in this chapter. Therefore, we introduced three methods to convert a continuous-time system to digital form. These topics help the reader to replace an analog system with its digital counterpart.

Problems

5.1 Consider the continuous-time system with transfer function

$$G(s) = \frac{20(s+5)}{s(s+4)(s+6)+25}$$

Using the Python control systems library
a) obtain digital version of the system using pole-zero matching, zero order hold equivalent, and bilinear transformation. Set $T_s = 0.05$ seconds for transformations.
b) plot unit step response of the obtained discrete systems.

5.2 Repeat Problem 5.1 when $T_s = 0.02$ seconds. What has changed?

5.3 Apply a sinusoidal signal to PA6 pin of the STM32 microcontroller using MicroPython. The sinusoidal signal should be between 1 and 2 V with frequency 10 Hz.

a) Implement the ADC operation by setting the sampling frequency to 500, 100, 20, and 5 Hz. Store 500 samples in an array. The quantization level for the operation should be 12 bits. Transfer the acquired data to PC and analyze the results.

b) Repeat the steps in part (a) by setting the quantization level to 8 bits. For this case, use the sampling frequency as 500 Hz. Comment on the effect of quantization level in the ADC operation.

5.4 Generate a sawtooth signal on the STM32 microcontroller and feed it through PA4 pin using MicroPython. The period of the signal should be 10 Hz. The amplitude of the signal should be between 1 and 2 V.

a) Implement the DAC operation by setting the sampling frequency to 100, 250, and 1000 Hz. Feed the generated data to output and observe it by an oscilloscope and analyze the results.

b) Repeat the steps in part (a) by setting the quantization level to 8 bits. For this case, use the sampling frequency as 1000 Hz. Comment on the effect of quantization level in the DAC operation.

5.5 Repeat Problem 5.3 using C language on Mbed Studio.

5.6 Repeat Problem 5.4 using C language on Mbed Studio.

6

Constructing Transfer Function of a System

The first step in controlling a system is to know how it behaves for different input signals. The input output relation obtained for this purpose is called transfer function of the system as explained in detail in Chapter 4. Afterward, control methods can be applied on it. This chapter focuses on constructing the transfer function of a system. To do so, we will use two fundamental methods as mathematical modeling and system identification. Mathematical modeling is a theoretical approach in which the system structure should be known in advance. On the other hand, system identification is a more general approach which can be applied to any system. Through these two approaches, we will obtain both the continuous- and discrete-time transfer functions for the system of interest. Here, we pick the simple RC filter as an electrical system to explain the concepts in the following sections. Besides, we pick the DC motor as a more advanced electro-mechanical system at the end of chapter application. We will benefit from the extracted transfer functions here throughout the book. The reader can also apply the methods introduced in this chapter to his or her system to understand its characteristics. Hence, suitable controllers can be designed for it as will be practiced in the following chapters.

6.1 Transfer Function from Mathematical Modeling

It is mandatory to know the relation between input and output of a system to analyze and design a controller for it properly. We can use mathematical modeling if the system and its components can be described this way. Here, the system is represented as a set of mathematical equations using its components. Then, Laplace transform can be used to obtain transfer function of the system if it is linear or can be linearized. Afterward, this transfer function can be converted to

Embedded Digital Control with Microcontrollers: Implementation with C and Python,
First Edition. Cem Ünsalan, Duygun E. Barkana, and H. Deniz Gürhan.
© 2021 The Institute of Electrical and Electronics Engineers, Inc. Published 2021 by John Wiley & Sons, Inc.
Companion website: www.wiley.com/go/Unsalan/Embedded_Digital_Control_with_Microcontrollers

digital form by the techniques introduced in Chapter 5. Therefore, we will start with providing the fundamental electrical and mechanical components, as well as their characteristics, in this section. Afterward, we will construct the differential equation representing the system based on these. Finally, we will obtain the digital transfer function from the constructed differential equation.

6.1.1 Fundamental Electrical and Mechanical Components

We can benefit from three fundamental components to represent any LTI electrical system. These are the resistor, capacitor, and inductor. These components can be described by the voltage, $e(t)$, and current, $i(t)$, relation they have as tabulated in Table 6.1. We will use these in constructing the differential equation representing the system.

As in electrical systems, we have three fundamental components to represent any LTI mechanical system. These are the damper, mass, and spring. Characteristics of these components can be described by the force, $f(t)$, and velocity, $v(t)$, relation they have as tabulated in Table 6.2.

As can be seen in Tables 6.1 and 6.2, there is a clear analogy between electrical and mechanical components. To do so, we should associate voltage with force and current with velocity. This may be of great help in analyzing electrical and mechanical systems. For more information on this topic, please see the book by Ogata (2009).

Table 6.1 Fundamental electrical components.

Component	Constant	Voltage–current relationship
Resistor	R	$e(t) = Ri(t)$
Capacitor	C	$i(t) = C\,de(t)/dt$
Inductor	L	$e(t) = L\,di(t)/dt$

Table 6.2 Fundamental mechanical components.

Component	Constant	Force–velocity relationship
Damper	B	$f(t) = Bv(t)$
Mass	M	$f(t) = M\,dv(t)/dt$
Spring	K	$v(t) = 1/K\,df(t)/dt$

6.1.2 Constructing the Differential Equation Representing the System

We can use the fundamental components and their connection in forming the system's differential equation. Here, Kirchhoff voltage law (KVL) and Kirchhoff current law (KCL) will be useful for electrical systems. Likewise, Newton's laws may be of help for mechanical systems. We will benefit from KVL and KCL in obtaining the differential equation of a first-order electrical system as resistive capacitive (RC) filter as an example. We provide it next.

Example 6.1 *(Simple RC filter, constructing the differential equation)*
Circuit diagram of the simple RC filter is in Figure 6.1. Let us take the input for this system as voltage represented by $x(t)$. Furthermore, let us take output as voltage on the capacitor represented as $y(t)$. Using KVL, we can form the differential equation relating input and output as $y(t) = x(t) - Ri(t)$. Using the current–voltage relationship of the capacitor tabulated in Table 6.1, we can obtain

$$y(t) = x(t) - RC\frac{dy(t)}{dt} \tag{6.1}$$

This is the differential equation representing the simple RC filter.

As can be seen in this example, input and output of the system is represented by a differential equation. Next, we will obtain transfer function of the system using this equation.

6.1.3 From Differential Equation to Transfer Function

Differential equation representing the system should be solved to obtain its output for a given input. This may not be easy for most input signals. Instead, we can represent the differential equation in complex domain by using Laplace transform. Then, we can obtain transfer function of the system in complex domain. Afterward, we can benefit from the methods introduced in Section 5.3 to represent it

Figure 6.1 Simple RC filter.

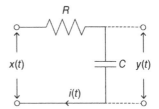

in digital form. To explain this method in more detail, we provide the RC filter example next.

Example 6.2   (Simple RC filter, constructing the transfer function)
When we apply Laplace transform to the differential equation in Eq. (6.1), we obtain $Y(s) = X(s) - sRCY(s)$. Here, $X(s)$ and $Y(s)$ represent Laplace transform of the input and output signals for the simple RC filter, respectively. Rearranging the representation, we can obtain

$$\frac{Y(s)}{X(s)} = \frac{\frac{1}{RC}}{s + \frac{1}{RC}} \tag{6.2}$$

where $\frac{Y(s)}{X(s)} = G(s)$ is the transfer function of the simple RC filter.

Let us pick the component values as $R = 10$ kΩ and $C = 100$ ηF. We can obtain the final transfer function of our simple RC filter using these values as

$$G(s) = \frac{1000}{s + 1000} \tag{6.3}$$

As we have transfer function of the system at hand in complex domain, we can convert it to time domain by the inverse Laplace transform (Oppenheim et al. 1997). Hence, we can obtain impulse response of the continuous-time system. For our case, we can benefit from the methods introduced in Chapter 5 to obtain digital representation of the system. Let us show how this can be done in the next example.

Example 6.3   (Simple RC filter, obtaining the digital transfer function)
We can use the bilinear transformation method to obtain transfer function of the simple RC filter in digital form. Assume that the sampling period for operation is taken as 0.0005 seconds. We can obtain the corresponding transfer function as

$$G(z) = \frac{0.2z + 0.2}{z - 0.6} \tag{6.4}$$

We can use this transfer function to analyze the system characteristics and design appropriate controllers for our simple RC filter.

6.2   Transfer Function from System Identification in Time Domain

Although mathematical modeling can be used in some applications, it depends on the differential equation representing the system. This may not be possible for some systems. For such cases, we can use system identification in which different kind of input signals are fed to the system. Corresponding output signals obtained from the system are recorded. Then, transfer function of the system is constructed using these input–output signal pairs.

Depending on input signals, we can categorize system identification into two groups as time and frequency domain based. In this section, we will only focus on system identification in time domain. In the next section, we will introduce the frequency domain approach.

6.2.1 Theoretical Background

System identification is basically an iterative trial and error approach. Here, the actual system, $g[n]$, is modeled by another system, $g_e[n]$, having a parametric representation. These two systems are shown in Figure 6.2.

In order to identify $G(z)$, we feed an input signal $x[n]$ to it and obtain the output $y[n]$ as in Figure 6.2(a). Then, the same input signal is applied to $G_e(z)$ and the output $y_e[n]$ is obtained as in Figure 6.2(b). Then, an error is defined based on the difference between $y[n]$ and $y_e[n]$. Our aim here is to minimize this error by adjusting the parameters of $G_e(z)$. We iteratively repeat this process by applying different input signals. After certain number of iterations, we expect to have an acceptable error. Hence, we can deduce that $G_e(z)$ represents $G(z)$ fairly well. In other words, the transfer function $G(z)$ is identified.

There are three different linear models to form $G_e(z)$. These are polynomial, transfer function, and state-space models. We will use the transfer function model throughout this book. To be more precise, $G_e(z)$ will be represented as an LTI system with numerator and denominator polynomials in z-domain. Parameters to be adjusted will be coefficients of these polynomials. We will benefit from MATLAB in system identification in time and frequency domains. Therefore, we do not provide the detailed theoretical explanation here. For more information on the theoretical background of system identification, please see the books by Söderström and Stoica (1989) and Ljung (1999).

6.2.2 The Procedure

While applying system identification in time domain, we should pay attention to four points to obtain good results. First, the chosen input signal must excite the system to be identified such that all system dynamics can be observed at output.

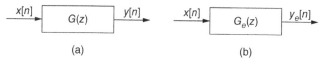

        (a)   (b)

Figure 6.2 Schematic representation of system identification. (a) Unknown system. (b) Estimated system.

Related to this, we should know how the system behaves under slow or fast rising/falling edges of a signal. Simple input signals, such as unit step and single frequency sine signal, are generally not sufficient to observe these behaviors. Second, the sampling period for the system to be identified must be chosen carefully to observe its dynamic response at output. As an example, if the system at hand responds to the impulse input within 0.1 second and the selected sampling period is one second, this dynamic response will most likely be missed. Therefore, the user should apply a unit step input to the system to be identified and record the response as output. Then, the sampling period should be selected as at most one tenth of the rise time. For more information on rise time, please see Chapter 7. Third, duration of the input signal must be chosen carefully to observe all important time constants of the system to be identified. Fourth, noise will affect the system identification operation. Hence, precautions should be taken at the data acquisition step for noise reduction.

Based on the mentioned points above, the most suitable input for system identification is the random signal. The main reason is that this signal contains all excitation properties for any system. However, the random signal should have a proper sampling period. Besides, its duration should be selected properly for the system to be identified. The sum of sinusoids signal is also another good option to be used in system identification. However, this signal should be formed such that sufficient number of sinusoids should be available in the signal. Besides, rectangular and step signals can be used for identifying low-order systems.

In order to identify a system, we should feed digital signals to it as input and record the output signals from it. The most suitable option for this operation is using the STM32 microcontroller. We provide detailed explanation for the setup of the microcontroller next. To note here, we will base our explanation on the simple RC filter. We will also handle the DC motor as the end of chapter application.

6.2.3 Data Acquisition by the STM32 Microcontroller

We provide block diagram of the general setup for system identification using the STM32 microcontroller in Figure 6.3. As can be seen in this figure, the system takes the input signal $x[n]$ as input and generates the output signal $y[n]$.

We should use ADC, DAC, and USART modules of the STM32 microcontroller in C language to obtain data for system identification. This setup has been explained in detail in Section 5.4. The same setup can be used here to obtain datasets for different input signals. The setup given in Section 5.4 was also used to apply step input to the RC filter from DAC, obtain output from ADC and send input–output pair to PC using USART module in MicroPython. Therefore, the same setup can be used here to obtain datasets for different input types in MicroPython.

Figure 6.3 Block diagram of the general setup for system identification in time domain.

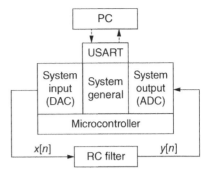

We can feed four different input signals to the system for time domain system identification. The first input is the unit step signal. The second input is the rectangular signal with 1 V amplitude, 0.5 V offset value, 2000 samples period, and 50% duty cycle. The third input is the sum of five sinusoids signal with frequencies 2, 20, 50, 100, and 200 Hz, each having amplitude of 0.1 V, offset value 0.2 V and phase 0 rad. The fourth input is the random signal with 1 V amplitude, 0.5 V offset value, and 100 sample random sample duration. Sampling period for all these signals is selected as 0.0005 seconds. Duration of each signal is selected as five seconds.

Each input should be applied to the system separately and corresponding outputs should be recorded. The Python script given in Section 4.7 can be used for this purpose. As a result, we should have four data pairs such as `RC_Step_Input.dat` and `RC_Step_Output.dat`.

6.2.4 System Identification in Time Domain by MATLAB

We use MATLAB system identification toolbox to obtain transfer function of a system in this book. Here, we will explain how this toolbox can be used for our simple RC filter. At this step, we should have the input–output signal pairs of the system acquired by the STM32 microcontroller. These should have been stored as separate files by the Python code given in the previous section. We should import these files to MATLAB first. To do so, please use the code in Listing 6.1. Do not forget to change the MATLAB working directory according to your Python workspace.

Listing 6.1: The MATLAB code for storing input–output data.

```
clear

fid=fopen('RC_Step_Input.dat','rt');
dummy = textscan(fid, '%f');
input_Step = dummy{1};
fclose(fid);

fid=fopen('RC_Step_Output.dat','rt');
dummy = textscan(fid, '%f');
```

```
output_Step = dummy{1};
fclose(fid);

fid=fopen('RC_Rect_Input.dat','rt');
dummy = textscan(fid, '%f');
input_Rect = dummy{1};
fclose(fid);

fid=fopen('RC_Rect_Output.dat','rt');
dummy = textscan(fid, '%f');
output_Rect = dummy{1};
fclose(fid);

fid=fopen('RC_SoS_Input.dat','rt');
dummy = textscan(fid, '%f');
input_SoS = dummy{1};
fclose(fid);

fid=fopen('RC_SoS_Output.dat','rt');
dummy = textscan(fid, '%f');
output_SoS = dummy{1};
fclose(fid);

fid=fopen('RC_Random_Input.dat','rt');
dummy = textscan(fid, '%f');
input_Random = dummy{1};
fclose(fid);

fid=fopen('RC_Random_Output.dat','rt');
dummy = textscan(fid, '%f');
output_Random = dummy{1};
fclose(fid);


Ts= 0.0005;

Step_data = iddata(output_Step,input_Step,Ts);
Rect_data = iddata(output_Rect,input_Rect,Ts);
SoS_data = iddata(output_SoS,input_SoS,Ts);
Random_data = iddata(output_Random,input_Random,Ts);
```

After the desired time domain data are imported, corresponding data objects should be created under MATLAB using the function iddata(y,x,Ts). Here, Ts is the sampling period, y is the output signal, and x is the input signal. Then, open the system identification toolbox by entering the command systemIdentification in the MATLAB command window. A GUI should pop up as in Figure 6.4.

From the opened GUI, time domain data objects can be added using the "Import data" drop-down menu. From this menu, click "Data object ..." and "Import data." A new window will be opened as in Figure 6.5. Here, we should enter the object name to be imported and click "Import." The imported dataset should be seen in one of the eight boxes at the left side as data icon. In Figure 6.5, the first data object Step_data import operation is provided. The same operation should be repeated for the remaining three objects Rect_data, SoS_data, and Random_data successively. After an object is imported, its time domain plot can be obtained. To do

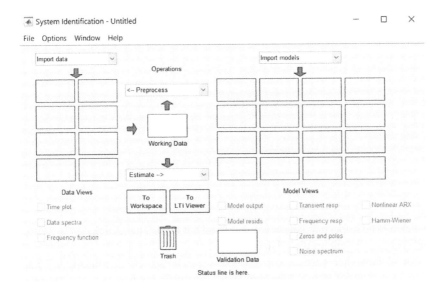

Figure 6.4 MATLAB system identification toolbox GUI. (Source: The MathWorks, Inc.)

Figure 6.5 Import data window. (Source: The MathWorks, Inc.)

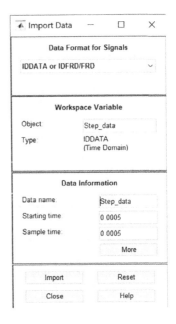

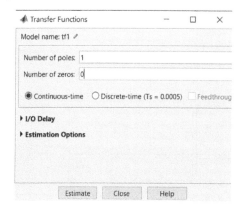

Figure 6.6 Transfer function window. (Source: The MathWorks, Inc.)

so, click on the imported dataset icon and make it bold. Then, click on the "Time Plot" checkbox to obtain the related plot.

After all data objects are imported, we are ready for system identification. To do so, drag one dataset to the "Working Data" box and one dataset to the "Validation Data" box. The dataset selected for working data is used for identifying the transfer function. To perform identification, click on the "Estimate →" dropdown menu and select "Transfer Function Models ..." The opening window will be as in Figure 6.6.

In Figure 6.6, the transfer function can be identified either in continuous- or discrete-time by selecting the related checkbox. Also, number of poles and zeros of the transfer function should be entered to the related boxes before the identification process. For our simple RC filter, please enter the number of poles as 1, number of zeros as 0, and select continuous-time transfer function from the related checkboxes. Finally, click on the "Estimate" button to start the identification process. As the process finalizes, the created transfer function object is placed in one of the 12 boxes at the right as data icon. The obtained transfer function and its properties can also be seen in the "Data/model Info" window by double-clicking on it as in Figure 6.7.

After the estimation process finishes, dataset selected for validation data can be applied to the estimated transfer function. Here, input of the validation dataset is fed to the transfer function, and output of this process is compared with output of validation dataset. This comparison can be also observed on a plot by clicking "Model output" checkbox. Also residual analysis can be performed by clicking "Model resids" checkbox.

After identifying the system, its transient response and pole-zero locations can be observed. To do so, please click on the estimated transfer function and make it bold. Then, click on "Transient resp" or "Zeros and Poles" box to obtain the related plots.

Figure 6.7 Data/model info window. (Source: The MathWorks, Inc.)

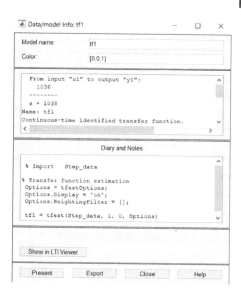

Table 6.3 Comparison of estimated transfer functions.

Input	Transfer function	Fit percentage (%)
Random signal	$G(s) = \dfrac{1054}{s + 1055}$	98.25
Sum of sinusoids	$G(s) = \dfrac{1057}{s + 1059}$	98.23
Rectangular signal	$G(s) = \dfrac{1048}{s + 1048}$	98.19
Unit step signal	$G(s) = \dfrac{1036}{s + 1038}$	98.18

For our simple RC filter, the obtained transfer functions and their fit percentages to validation data (output of random input signal) are tabulated in Table 6.3. As can be seen in this table, the fit percentage for all estimations are fairly high. This is because the RC filter is a first-order system. Moreover, the highest fit percentage is obtained when the input is selected as random signal as expected. The sum of sinusoids input signal also has a fairly good fit percentage. Both transfer functions are close to the one calculated by mathematical modeling given in Eq. (6.3). The time domain system identification results obtained by the rectangular and unit-step input signals are also acceptable.

6.3 Transfer Function from System Identification in Frequency Domain

System identification can be performed in frequency domain as well besides time domain. In fact, the only difference between the time and frequency domain approaches is the used input signals and the selected candidate transfer function representations. Besides, the procedure to be followed in both approaches is the same. Since we introduced the time domain approach in detail in Section 6.2, we will only highlight the differences for the frequency domain based system identification here. Besides, we will direct the reader to the related parts in time domain system identification.

6.3.1 Theoretical Background

In order to perform system identification in frequency domain, we should represent the system as such. In other words, the transfer function of the system to be identified should be represented as $G(j\omega) = |G(j\omega)|e^{j\angle G(j\omega)}$. Then, the magnitude, $|G(j\omega)|$, and phase, $\angle G(j\omega)$, terms for each frequency should be calculated from the given input–output signal pair. For more information on the theory of system identification in frequency domain, please see the book by Pintelon and Schoukens (2001).

6.3.2 The Procedure

Since system identification in frequency domain requires calculating the magnitude, $|G(j\omega)|$, and phase, $\angle G(j\omega)$, terms for each frequency, appropriate input signals should be fed to the system. We have two options here. The first one is applying sinusoidal signals with different frequency to the system as input. The system response to these are recorded as output. Here, the frequency resolution of input signals must be selected carefully. This resolution should be small enough such that it contains all important frequency information. On the other hand, the frequency resolution should not be too small in order to prevent obtaining unnecessary experimental data. The second option is applying a high bandwidth, long duration sweep signal to the system.

For our RC filter, we select the input signal as a sweep signal with start and stop frequency values as 1 and 400 Hz, respectively. Amplitude of the signal is set as 0.5 V. The offset value is 1 V. Duration of the signal is set as 60 seconds.

The input signal and received output are stored in MATLAB. Afterward, fast Fourier transform is applied to them to obtain the corresponding frequency data. The MATLAB code used to perform all these operations is given in Listing 6.2.

Listing 6.2: The MATLAB code for obtaining frequency domain dataset.

```
fid=fopen('RC_Sweep_Input.dat','rt');
dummy = textscan(fid, '%f');
input_Sweep = dummy{1};
fclose(fid);

fid=fopen('RC_Sweep_Output.dat','rt');
dummy = textscan(fid, '%f');
output_Sweep = dummy{1};
fclose(fid);

Ts= 0.0005;

Sweep_data = iddata(output_Sweep,input_Sweep,Ts);
Frequency_data = fft(Sweep_data);
```

6.3.3 System Identification in Frequency Domain by MATLAB

The MATLAB interface introduced in Section 6.2 can also be used in system identification in frequency domain. We apply system identification in frequency domain with the data object formed in Listing 6.2. This way, we obtain the final transfer function for our RC filter as $G(s) = \frac{1053}{s+1055}$. This transfer function is similar to the ones obtained in time-domain based method. As we validate the transfer function with the random signal used in the previous section, we obtain the fit value as 95.24%. This also indicates that the system identification in frequency domain works as expected.

6.4 Application: Obtaining Transfer Function of the DC Motor

Here, we will obtain transfer function of our DC motor. To do so, we will start with mathematical modeling. Then, we will use system identification in time and frequency domains.

6.4.1 Mathematical Modeling

The DC motor can be represented as in Figure 6.8. In this figure, $e_a(t)$ is the applied armature voltage; $i_a(t)$ is the armature current; R_a is the armature resistance; L_a is

Figure 6.8 Representing the DC motor.

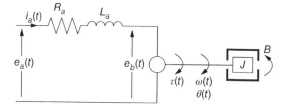

the armature inductance; $e_b(t)$ is the back electromotive force (emf) voltage; $\tau(t)$ is the motor torque; $\omega(t)$ is the angular velocity; $\theta(t)$ is the angular position; J is the rotor moment of inertia; and B is the viscous friction coefficient.

Mechanical and electrical motor equations are required to extract transfer function of the motor. Therefore, electrical equation of the motor is obtained by applying KVL to the armature as

$$e_a(t) = i_a(t)R_a + L_a\frac{di_a(t)}{dt} + e_b(t) \tag{6.5}$$

The relationship between motor torque and armature current is $\tau(t) = K_r i_a(t)$, where K_r represents the torque constant. The relation between back emf voltage and angular velocity is $e_b(t) = K_b\omega(t)$, where K_b represents the back emf constant. The torque equation with no load is

$$J\frac{d\omega(t)}{dt} = \tau(t) - B\omega(t) \tag{6.6}$$

When Laplace transform is applied to these equations, we obtain

$$E_a(s) - E_b(s) = I_a(s)(R_a + sL_a) \tag{6.7}$$

$$\tau(s) = K_r I_a(s) \tag{6.8}$$

$$E_b(s) = K_b\omega(s) \tag{6.9}$$

$$\tau(s) = (sJ + B)\omega(s) \tag{6.10}$$

Substituting Eqs. (6.8) and (6.9) into Eqs. (6.7) and (6.10), we obtain

$$E_a(s) - K_b\omega(s) = I_a(s)(sL_a + R_a) \tag{6.11}$$

$$(sJ + B)\omega(s) = K_r I_a(s) \tag{6.12}$$

When Eqs. (6.11) and (6.12) are merged, transfer function of the DC motor can be obtained as

$$\frac{\omega(s)}{E_a(s)} = \frac{K_r}{(sL_a + R_a)(sJ + B) + K_r K_b} \tag{6.13}$$

Based on the transfer function in Eq. (6.13), block diagram of our DC motor can be represented as in Figure 6.9.

To note here, $\omega(s)$ in Eq. (6.13) is represented by rad/s. However, we read speed from the encoder in rpm form. Hence, it should be converted to rpm form by the equality 1 rad/s = 9.5493 rpm. Thus, Eq. (6.13) can be rewritten as

$$\frac{\omega(s)}{E_a(s)} = \frac{9.5493K_r}{(sL_a + R_a)(sJ + B) + K_r K_b} \tag{6.14}$$

Actual parameter values R_a, L_a, K_b, K_r, B, and J must be known to construct the overall transfer function in Eq. (6.14). If these values are not tabulated in the

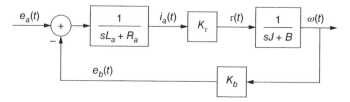

Figure 6.9 Block diagram of the DC motor.

datasheet of the motor, then they should be obtained experimentally. To do so, we should apply the following procedure.

When rotation of the motor is blocked and system reaches steady state, Eq. (6.5) becomes $R_a = \frac{e_a}{i_a}$. The e_a voltage is swept and corresponding i_a values are recorded. It is possible to find R_a by averaging the recorded data values. The L_a value can be obtained in two ways. As the first option, it can be measured by using an LCR meter. As the second option, the reader can connect the AC source to the motor terminals and measure AC voltage and current. Then, the motor impedance, the vector sum of motor resistance and reactance, can be calculated. Hence, L_a can be calculated. When the system reaches steady state, K_b can be obtained using $K_b = \frac{e_a - i_a R_a}{\omega}$. When R_a is known, e_a and i_a can be measured using a multimeter. ω can be obtained with the help of a tachometer. Or, it can be calculated by looking at the encoder output from an oscilloscope. K_t can be taken equal to K_b if the motor speed is represented as rad/s. To note here, we read speed from the encoder in rpm form. Hence, it should be converted to rad/s form by the equality 1 rad/s = 9.5493 rpm. When the system reaches steady state, Eq. (6.10) turns to $\tau = B\omega$. Then, $B = \frac{K_t i_a}{\omega}$. K_t, i_a, and ω are already obtained. Therefore, B can be calculated. In order to find J, a step input should be applied to the unloaded motor. The mechanical time constant is measured which is the amount of time for the motor to rise to 63.2% of its final velocity. Then, J can be calculated using $\tau_m = \frac{RJ}{K_b K_t}$.

For our motor, we applied the above procedure and obtained the parameter values as $R_a = 3.2\ \Omega$, $L_a = 8.2\ \text{mH}$, $K_b = 0.85\ \text{Vs/rad}$, $K_t = 0.85\ \text{Nm/A}$, $B = 0.016\ \text{Nms/rad}$, and $J = 0.0059\ \text{Nms}^2/\text{rad}$. Based on these parameters, we can obtain the transfer function in Eq. (6.14) as

$$G(s) = \frac{\omega(s)}{E_a(s)} = \frac{167773.98}{s^2 + 392.96s + 15992.15} \tag{6.15}$$

This transfer can be represented in digital form using bilinear transformation as

$$G(z) = \frac{0.0095z^2 + 0.0191z + 0.0095}{z^2 - 1.8176z + 0.8213} \tag{6.16}$$

where, the sampling period is set as 0.0005 seconds. Poles of the system are located at 0.9772 and 0.8404 which are inside the unit circle. Thus, the system is stable.

6.4.2 System Identification in Time Domain

We can apply the procedure in Section 6.2 to identify transfer function of the DC motor in time domain. The first and most important step for this operation is selecting the sampling frequency. It should be selected high enough to observe all dynamic responses of the system. Therefore, we should apply a unit-step input to the motor and record the output speed signal. Then, we should calculate the rise time from it. The sampling period must be selected at least 10 times smaller than this value. We applied this procedure to our DC motor and obtained the rise time to be approximately 0.05 seconds. Therefore, we set the sampling frequency to be 2 kHz ($T_s = 0.0005$ seconds) in our case.

We provide the general setup to identify transfer function of the DC motor in Figure 6.10. As can be seen in this figure, the DC motor takes the input signal $x[n]$. This signal is generated by the PWM module of the microcontroller as voltage. This PWM signal is amplified by the motor driver to 0–12 V range.

Output speed of the motor, $y[n]$, is observed by the encoder of the motor which generates two rectangular signals as $s1[n]$ and $s2[n]$. The encoder mode of the timer module of the microcontroller is used to calculate $y[n]$ from these two signals. Both input and output signals are then fed to PC using the Python interface as explained in Section 6.2. Then, these signals are processed under MATLAB for the system identification operation.

We fed the random signal as input to the DC motor. Here, the duration of the signal is selected as 20 seconds for obtaining valid measurements. As the input–output signal pair is obtained in the microcontroller, it is sent to PC through the Python interface there. Then, time domain data object is created using the MATLAB function `iddata(y,u,Ts)` for this input–output signal pair.

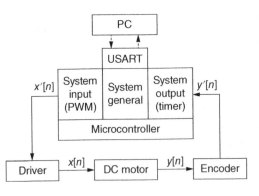

Figure 6.10 Block diagram of the general setup for the DC motor by system identification.

We already know that the system to be identified should have two poles and no zero from mathematical modeling. Hence, the system identification toolbox under MATLAB is executed for the created time domain data object to obtain second-order transfer function with no zero. The obtained transfer function is

$$G(s) = \frac{\omega(s)}{E_a(s)} = \frac{165702.47}{s^2 + 390.20s + 15701} \tag{6.17}$$

where the fit to the validation data is 88% in time domain system identification. As we compare the transfer function obtained by mathematical modeling, given in Eq. (6.15), and the transfer function in Eq. (6.17), we can see that they have similar coefficients. Hence, system identification in time domain is justified for this case.

We can represent the transfer function in Eq. (6.17) in digital form using bilinear transformation and obtain

$$G(z) = \frac{0.0094z^2 + 0.0189z + 0.0094}{z^2 - 1.8188z + 0.8224} \tag{6.18}$$

where, the sampling period is set to 0.0005 seconds. As can be seen here, poles of the system are located at 0.9775 and 0.8413. Poles obtained in mathematical modeling were 0.9772 and 0.8404, respectively. Hence, we can also observe that the poles in z-domain obtained by mathematical modeling and system identification in time domain are close to each other.

6.4.3 System Identification in Frequency Domain

We can obtain transfer function of the DC motor by system identification in frequency domain as well. To do so, we applied 160 sinusoidal signals to the system from 1 to 160 Hz with 1 Hz frequency steps. Sampling time for all these signals is selected as 0.0005 seconds. Duration of the signals is selected as two seconds. During operation, each input signal is applied to the system and corresponding output is recorded separately. Afterward, each output signal is fit to a separate sinusoidal signal to eliminate noise term on it. Then, amplitude and phase shifts for these 160 points are calculated using input and fitted output sinusoidal signals. Based on this, frequency domain data object is created in MATLAB using the function `idfrd(Response, Freq, Ts)`. Finally, the system identification toolbox is run for created frequency domain data object to obtain the second-order transfer function with no zeros. The obtained transfer function is

$$\frac{\omega(s)}{E_a(s)} = G(s) = \frac{225232.30}{s^2 + 548.96s + 21341.70} \tag{6.19}$$

where the fit to the validation data is 80% in frequency domain system identification. As we compare this transfer function by the previously obtained two transfer functions, we can see that they differ in coefficient values.

We can represent the transfer function in Eq. (6.19) in digital form using bilinear transformation and obtain

$$G(z) = \frac{0.0124z^2 + 0.0246z + 0.0124}{z^2 - 1.7542z + 0.7589} \tag{6.20}$$

where, the sampling period is set to 0.0005 seconds. As can be seen here, poles of the system are located at 0.9792 and 0.7751. These poles and the ones obtained by previous two methods are different as well. We should mention one possible reason for this difference. Amplitude of the output signal became too small after 160 Hz. Hence, it could not be fitted to a sinusoidal correctly because of fluctuations at output. Therefore, all frequency range could not be calculated. As a result, high frequency pole of the system could not be estimated correctly.

6.5 Summary

Representing a physical system by its transfer function is mandatory while designing a controller for it. Therefore, this chapter introduced three methods for extracting the transfer function of a system. During operation, we assumed an LTI system and acted accordingly. The first method we considered in the transfer function extraction operation was based on mathematical modeling. If building blocks of a physical system and differential equation representing the overall system is at hand, this method can be used with confidence. Besides, it does not require any measurements on the system. The second and third transfer function extraction methods depend on system identification in time and frequency domains, respectively. These assume the system to be a black box. They obtain transfer function of the system based on its input and output signals. The STM32 microcontroller is a perfect candidate for obtaining the input and output signals from the actual system. We applied all three methods on the simple RC filter and DC motor throughout the chapter. The reader can apply the same setup to his or her physical system by following the procedures given in this chapter. Therefore, the physical system can be represented by its transfer function. This leads to digital controller design to be considered in the following chapters.

Problems

6.1 What is system identification, why is it used, and what are its types?

6.2 Reconsider the simple RC filter given in Figure 6.1, with $R = 10\,k\Omega$ and $C = 100\,\eta F$,

a) apply system identification in time domain with the sampling frequency set to 50 Hz and 100 Hz.

b) apply system identification in frequency domain with the sampling frequency set to 50 and 100 Hz.

c) what has changed in the characteristics of identified systems?

6.3 Reconsider the simple RC filter given in Figure 6.1, now with $R = 100 \, k\Omega$ and $C = 1 \, \eta F$,

a) construct transfer function of this new system by mathematical modeling.

b) pick an appropriate sampling frequency for operation. Apply system identification in time domain with the selected sampling frequency.

c) apply system identification in frequency domain with the selected sampling frequency.

d) compare the obtained results.

7

Transfer Function Based Control System Analysis

A system has its inherent characteristics. These may not satisfy the required target performance criteria. We can make sure that these are satisfied by adding a controller to the system. Hence, we should first understand what are the possible performance criteria. To do so, we will introduce methods to analyze the system in time domain, frequency domain, and complex plane. These three domains are related. However, they provide valuable information on the same system from different perspectives. We will next explore open- and closed-loop controllers as well as their effects on system performance. Although controller design via transfer functions will be covered in Chapter 8, we believe the system analysis should be understood first. Therefore, we will cover fundamental concepts in control system analysis by transfer functions in this chapter. To be more specific, we will analyze digital systems constructed by open- and closed-loop control methods. We will also provide the corresponding Python functions for analysis methods. We will provide performance analysis of the actual DC motor as the end of chapter application.

7.1 Analyzing System Performance

We can analyze the performance of a system in three domains as time, frequency, and complex plane. We will introduce performance criteria for these in this section. Then, we will analyze the effect of open- and closed-loop controllers in time domain, frequency domain, and complex plane. We will also design controllers based on target performance criteria in these domains in Chapter 8. Therefore, let us start with the time domain analysis.

7.1.1 Time Domain Analysis

We can obtain valuable information from a system in time domain when a specific input signal is applied to it. The unit step signal is generally selected for this

Embedded Digital Control with Microcontrollers: Implementation with C and Python,
First Edition. Cem Ünsalan, Duygun E. Barkana, and H. Deniz Gürhan.
© 2021 The Institute of Electrical and Electronics Engineers, Inc. Published 2021 by John Wiley & Sons, Inc.
Companion website: www.wiley.com/go/Unsalan/Embedded_Digital_Control_with_Microcontrollers

purpose since it represents a significant and abrupt change in input. As this signal is fed to the system, we can observe its effect in two perspectives as transient response and steady-state error.

Assume that we have a DC motor at hand and apply unit step input to it (as a control signal) to change its speed to the desired final value. As input is applied to the motor, its speed undergoes a gradual change from initial to final value. The transient response deals with this gradual change period. Afterward, the motor reaches a constant speed. The difference between this and desired final value is defined as the steady-state error.

7.1.1.1 Transient Response

A system (such as DC motor) cannot change its output to a given input in zero time due to its physical characteristics. Output of the system during this change phase is called transient response of the system. Assume that we have a DC motor at hand and apply control signal to change its speed. Here, a slow response makes the DC motor to reach the desired speed in a longer time. Whereas an excessively rapid response makes the DC motor to go high speed in a shorter time. However, this fast action may cause a permanent damage to the motor. Thus, it is mandatory to analyze transient response of the system carefully.

Unit step input is generally selected for obtaining transient response of the system since this input signal represents a significant and abrupt change. Moreover, the system is assumed to be initially at rest. Hence, it starts responding to the unit step input by zero initial conditions. The unit step response can be thought as pushing start button of the DC motor. This is actually a control signal. It is important to know how the DC motor responds to this sudden input. In general, we are interested in knowing how low or high output of the system goes (as undershoot and overshoot) before it settles down and how fast it settles.

In order to grasp the transient response, let us first find the step response of the DC motor with transfer function $G(z)$ provided by Eq. (6.18). We can benefit from the MicroPython control systems library for this purpose. Let us consider this by the following example.

Example 7.1 *Step response of the DC motor*

From Eq. (6.18), we know that transfer function of the DC motor is

$$G(z) = \frac{0.0094z^2 + 0.0189z + 0.0094}{z^2 - 1.8188z + 0.8224} \tag{7.1}$$

We can use the Python code in Listing 7.1 on PC to obtain step response of the DC motor.

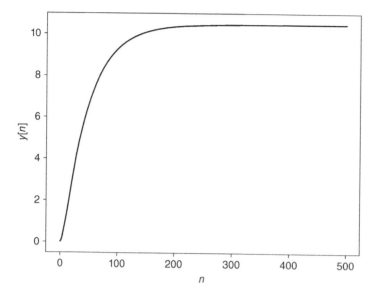

Figure 7.1 Step response of the DC motor.

Listing 7.1: Obtaining step response of the DC motor via MicroPython control systems library.

```
import mpcontrol
import mpcontrolPC

# System
Ts = 1
num = [0.00943, 0.01886, 0.00943]
den = [1, -1.8188, 0.8224]
g = mpcontrolPC.tf(num, den, Ts)

#Step input signal
N = 500
x = mpcontrol.step_signal(N, 1)

#Calculate the output signal
y = mpcontrol.lsim(g, x)
```

We can then plot the obtained output signal using the `matplotlib` library. The plot will be as in Figure 7.1. As can be seen in this figure, the response cannot reach the desired output value (as unit step).

Although it is possible to plot the step response of a given system and analyze the result visually, this is not feasible. Besides, it may not be possible to set controller design criteria (to be explored in the next chapter) just by the step response plot. Hence, we should summarize transient response of the system by some parameters or measures. Damping ratio, ζ, and natural frequency, ω_n, are the first two candidate parameters for this purpose.

If poles of a second-order system are complex as $z = re^{\pm j\theta}$, then ζ and ω_n can be calculated as

$$\zeta = \frac{-\ln r}{\sqrt{\ln^2 r + \theta^2}} \tag{7.2}$$

$$\omega_n = \frac{1}{T_s} \sqrt{\ln^2 r + \theta^2} \tag{7.3}$$

For higher-order systems, there will be ζ and ω_n associated with each complex pole pair or real pole of the transfer function. Here, we can take ζ and ω_n of the system as the ones obtained from its dominant poles. The dominant pole of a system is defined as the one inside and closest to the unit circle.

We can also calculate ζ and ω_n of a given system using the function tf_system.damp() within the MicroPython control systems library. This function originates from its MATLAB counterpart. Let us obtain ζ and ω_n of the DC motor considered in the previous example.

Example 7.2 *Damping ratio and natural frequency of the DC motor*
We can calculate ζ and ω_n of the DC motor, with the transfer function in Eq. (6.18), using the Python code in Listing 7.2 on PC. As we execute this code, we can see that the system has two poles at 0.8415 and 0.9773. We will obtain $\zeta = 1$. For the poles at 0.8415 and 0.9773, we will have $\omega_n = 345.1030$ rad/sec and $\omega_n = 45.9541$ rad/sec, respectively.

Listing 7.2: Calculating the damping ratio and natural frequency of the DC motor via MicroPython control systems library.

```
import mpcontrolPC

# System
Ts=0.0005
num = [0.00943, 0.01886, 0.00943]
den = [1, -1.8188, 0.8224]
g = mpcontrolPC.tf(num, den, Ts)

g.damp()
```

We can classify a system by looking at its ζ value which depends on the location of dominant poles of the system (Phillips et al. 2015). If ζ has a high value (such as 0.8), then the system is called overdamped. This means that, response of the system will not exceed the desired output value at any time instant. In other words, there will be no overshoot. At the same time, the system response will never reach the desired output value. If ζ has a small value, then the system is called underdamped. For this case, output of the system will oscillate with a decaying envelope around its final value. The decay rate of the envelope will be proportional to ζ such that when it is small (such as 0.1), the decay rate will be small as well. When ζ has a slightly larger value (such as 0.4), the decay rate will be large. Hence, the oscillation at output will be minimum.

Let us reconsider Example 7.2. As can be seen there, the DC motor has $\zeta = 1$. For this value, we expect the system will not have an oscillatory output to a given unit step input. Moreover, we do not expect the system to reach the desired output value. In fact, these two conditions can be observed in Figure 7.1.

We can summarize step response of the system by simple measures besides ζ. These are rise time, percent overshoot, peak time, and settling time.

Rise time, T_{rt}, represents the time required for the system response to go from 10% to 90% of its final value. Therefore, it is a measure of responsiveness of the system. In other words, the smaller T_{rt}, the more responsive is the system. We can associate T_{rt} and ζ as follows. When ζ is small, T_{rt} becomes smaller. However, there will be overshoot at output.

Percent overshoot, M_p, is the difference between maximum peak and final value of the unit step response. Related to this, peak time, T_{pt}, is when the maximum peak occurs. The amount of overshoot depends on ζ such that it becomes more severe when ζ is decreased. The system will have a moderate overshoot when there is medium ζ.

Settling time, T_{st}, is the time required for unit step response to reach and stay within 2% or 5% of its final value. The system with high ζ will respond slower to a given input signal compared to the system with medium ζ. Hence, it will take longer for the response to settle when ζ is high.

There are also supplementary measures besides the ones explained above. These are percent undershoot, settling min., settling max., and peak values. The percent undershoot is the ratio of the minimum system response (below zero) with its final value. If the minimum system response does not fall below zero, then this value is taken as zero. Settling min. represents the minimum system response after it reaches the rise time. Settling max. represents the maximum system response after it reaches the rise time. Finally, peak represents the maximum system response value.

We can obtain the transient response measures T_{rt}, M_p, T_{pt}, T_{st}, and supplementary measures for any system using the function stepinfo() in the MicroPython control systems library on PC. Let us consider the usage of this function on the DC motor considered in the previous example.

Example 7.3 *Transient response measures of the DC motor*

We can add the code line g.stepinfo() to the Python code in Listing 7.2 to obtain transient response measures of the DC motor. As we execute this code, we obtain $T_{rt} = 0.0485$ sec, $M_p = 0\%$, $T_{pt} = 0.4345$ sec, and $T_{st} = 0.088$ sec. These values can also be observed in Figure 7.1 when the figure index is multiplied by sampling time $T_s = 0.0005$ sec.

7.1.1.2 Steady-State Error

If the system is stable, we expect its output to reach its final (steady-state) value after the transient response. As an example, we expect the DC motor to reach a constant speed after its transient response. However, this value may be different from the desired output. The difference between these two values is called steady-state error, denoted by e_{ss}. If the system responds to the given input as expected, we should have $e_{ss} = 0$. In other words, the system should reach the desired output for the given input.

We can also obtain e_{ss} for any system using the function `ss_error()` in the MicroPython control systems library on PC. Let us use this function on the DC motor next.

Example 7.4 *Steady-state error of the DC motor*

As we add the code line `g.ss_error()` to the Python code in Listing 7.2, we can obtain steady state error of the DC motor. As we execute this code, we observe $e_{ss} = -9.4778$. This value is large and not acceptable. Therefore, we will use a controller to decrease it in the following sections.

7.1.2 Frequency Domain Analysis

Frequency domain provides valuable information on the general system characteristics. Therefore, we should obtain frequency response of the system, denoted by $G(j\omega)$. In theory, we can obtain $G(j\omega)$ by replacing z by $j\omega$ in $G(z)$. In practice, we can obtain $G(j\omega)$ by feeding sinusoidal signals with varying frequency to the system and observe magnitude and phase of the output signals as we did in Section 6.3. As a reminder, any LTI system will output a sinusoidal signal when its input is sinusoidal. Only the magnitude and phase terms will change due to frequency response of the system. We can analyze $G(j\omega)$ using Bode plot.

Bode plot is the graphical representation of $G(j\omega)$. It can also be used in controller design which is the subject of next chapter. $G(j\omega)$ is a complex function of frequency ω. Hence, it can be represented in polar coordinates in terms of magnitude, $|G(j\omega)|$, and phase, $\angle G(j\omega)$, as $G(j\omega) = |G(j\omega)|e^{j\angle G(j\omega)}$. Hence, Bode plot will have two parts as magnitude vs. frequency and phase vs. frequency.

The magnitude plot, $20\log|G(j\omega)|$, is expressed in decibel (dB) (assuming that the input signal has amplitude one), and the frequency is plotted in logarithmic scale. Hence, it becomes possible to examine the relationship between the magnitude and frequency over a large range. In the magnitude plot, any value above 0 dB means that the system is amplifying its input for that frequency. Hence, there is a gain in that frequency. To note here, 0 dB corresponds to log(1) which can happen when amplitude of the input signal to the system is equal to its output for the given frequency. Likewise, any magnitude value below 0 dB means that the system is attenuating its input for that frequency.

The phase plot, $\angle G(j\omega)$, indicates the phase shift introduced by the system for any frequency value. Note that, the key point in Bode plot is not only obtaining the magnitude and phase values wrt frequency, it also leads to better understanding of system behavior and its stability.

We can use the function `transfer_function.bode()` in the MicroPython control systems library on PC to obtain Bode plot of a system. Let us consider the DC motor and obtain its Bode plot using this function.

Example 7.5 *Bode plot of the DC motor*
We can use the Python code in Listing 7.3 on PC to obtain Bode plot of the DC motor.

Listing 7.3: Obtaining Bode plot of the DC motor via MicroPython control systems library.

```
import mpcontrolPC

#System
Ts = 0.0005
num = [0.00943, 0.01886, 0.00943]
den = [1, -1.8188, 0.8224]

g = mpcontrolPC.tf(num,den,Ts)

#Bode plot of the system
[mag, phase, freq] = mpcontrolPC.bode(g,1000)
```

As we execute the code in Listing 7.3, we obtain Bode plot of the system both in magnitude, phase, and frequency terms as `mag`, `phase`, and `freq` for 1000 elements. We can then use the `matplotlib` library to plot the magnitude and phase terms in Bode plot in Figures 7.2(a) and 7.2(b), respectively.

There are standard terminologies used to describe characteristics of a system based on its frequency response, hence, Bode plot. The first definition is DC gain which is the magnitude value at $\omega = 0$ Hz. In other words, DC gain is defined as $20 \log |G(0)|$. The second definition is the natural frequency, ω_n, introduced in time domain analysis. In magnitude vs. frequency plot, we can locate ω_n as the frequency in which the peak magnitude occurs. The third definition is bandwidth of the system. This is the frequency, ω_b, in which the magnitude value falls to -3 dB. Bandwidth of the system is commonly used as a measure of the upper limit in which system dynamics can be controlled. Any input signal having frequency above ω_b will be severely attenuated by the system.

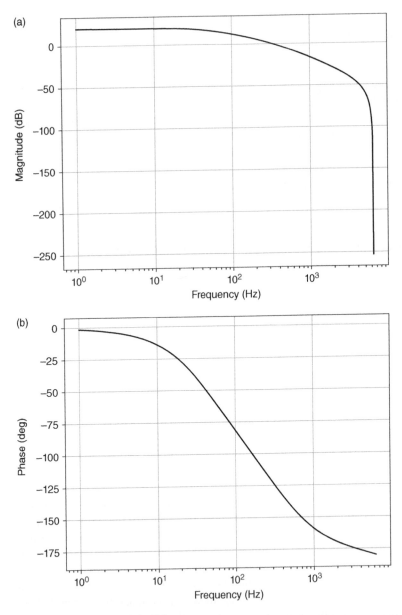

Figure 7.2 Bode plot of the DC motor.

There are also two terms defined in Bode plot as gain margin (GM) and phase margin (PM). These are used to describe stability of the system (Phillips et al. 2015). GM is defined as the amount of gain variation over the gain crossover frequency that will cause the system to become unstable. The gain crossover frequency, ω_{cg}, is where the GM is measured when the phase is at $-180°$. If GM is too small, then the system may become unstable. We can also define GM as the amount of gain allowed before the system becomes unstable. PM is defined as the amount of phase variation over the phase crossover frequency that will cause the system become unstable (Phillips et al. 2015). The phase crossover frequency, ω_{cp}, is where the PM is measured when the magnitude is at 0 dB.

PM is related to ζ. Small PM means small ζ and large overshoot, but fast response to input signal (as described in Section 7.1.1). The higher the PM, the more stable is the system. Higher PM means large ζ and less overshoot, but slow response to input signal (as described in Section 7.1.1).

Although it is possible to obtain the mentioned frequency response values of a given system mathematically, we can benefit from the function `margin()` in MicroPython control systems library on PC. Let us consider the DC motor and obtain the frequency response values.

Example 7.6 *Frequency response values of the DC motor*
We can use the Python code in Listing 7.4 on PC to obtain the frequency response values of the DC motor. As we execute this code, we obtain $GM = 313.071$ dB, $PM = 53.094°$, $\omega_{cg} = 6283.190$ rad/sec, and $\omega_{cp} = 338.802$ rad/sec. Since GM is larger than 3 dB, the system is stable.

Listing 7.4: Calculating frequency response values of the DC motor via MicroPython control systems library.

```
import mpcontrolPC

#System
Ts = 0.0005
num = [0.00943, 0.01886, 0.00943]
den = [1, -1.8188, 0.8224]

g = mpcontrolPC.tf(num,den,Ts)

#GM, PM, Crossover frequency values of the system
mpcontrolPC.margin(g)
```

7.1.3 Complex Plane Analysis

We can also analyze system characteristics in complex plane besides time and frequency domains. One possible option here is the root-locus plot. The other option is the Nyquist plot. We will evaluate both methods in this section.

7.1.3.1 Root-Locus Plot

Root-locus is a graphical technique used to determine the location of system poles based on a dependent variable. Although it is extensively used in the analysis of closed-loop systems (to be explored in the following sections), we can obtain the root-locus plot of a system as long as roots of its denominator polynomial can be represented as $1 + KG(z) = 0$. Based on this definition, root locations of the system can be plotted with respect to K. Hence, we can check whether the system is stable or not by looking at the pole locations wrt the unit circle for any K value. We will explain how the denominator of a system becomes as $1 + KG(z)$ in Section 7.3, when we deal with closed-loop control. Root-locus plot will be extensively used while designing closed-loop control systems in the next chapter as well.

Root-locus plot can be constructed based on its well-defined rules. However, it is not feasible to do so. Instead, we can benefit from the function `rlocus()` in Python control systems library on PC. Let us consider the DC motor example here.

Example 7.7 *Root-locus plot of the DC motor*
We can use the Python code in Listing 7.5 on PC to obtain root-locus plot of the DC motor.

Listing 7.5: Obtaining root-locus plot of the DC motor via Python control systems library.

```python
import control
import control.matlab

#System
Ts = 0.0005
num = [0.00943, 0.01886, 0.00943]
den = [1, -1.8188, 0.8224]

g = control.tf(num,den,Ts)

#Root locus plot of the system
control.matlab.rlocus(g)

from matplotlib import pyplot as plt

plt.ylim(-1,1)
plt.xlim(-1.5,1)
plt.grid()
plt.show()
```

We can examine how poles of the closed-loop DC motor system change when K changes using the root-locus plot given in Figure 7.3. It is possible for the designer to select any of these closed-loop poles for a specific K value that satisfies the desired T_{rt}, M_p, T_{pt}, T_{st}, and e_{ss} values.

7.1.3.2 Nyquist Plot

Nyquist plot is constructed using the magnitude and phase parts of Bode plot of a given system. As this plot is obtained, it relates stability of a closed-loop system

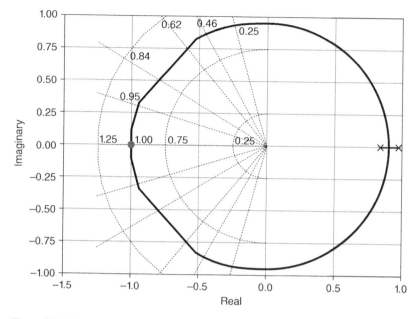

Figure 7.3 Root-locus plot of the DC motor with a controller added to it.

to its open-loop frequency response and pole locations. Thus, the open-loop system frequency response provides information about the stability of the closed-loop system. This concept is similar to the root-locus plot in which the open-loop poles and zeros of the system are used to deduce transient response and stability of the corresponding closed-loop system. The only difference of the Nyquist plot is that, we obtain the transient-response and stability information of the closed-loop system in frequency domain. Thus, it can be said that Nyquist plot is an alternative approach to root-locus. GM and PM can also be found by Nyquist plot. Hence, it can be used to determine the stability of the closed-loop system.

We can use the function `nyquist()` in Python control systems library to construct Nyquist plot of a given system on PC. Let us show how this can be done on the DC motor example.

Example 7.8 *Nyquist plot of the DC motor*
We can use the Python code in Listing 7.6 on PC to obtain Nyquist plot of the DC motor.

Listing 7.6: Obtaining Nyquist plot of the DC motor via Python control systems library.

```
import control
import math
```

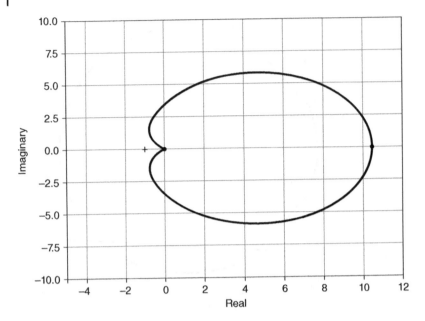

Figure 7.4 Nyquist plot of the DC motor.

```
import numpy

#System
Ts = 0.0005
num = [0.00943, 0.01886, 0.00943]
den = [1, -1.8188, 0.8224]

g = control.TransferFunction(num, den, Ts)

f_nyquist = 2*math.pi/Ts/2
k_nyquist = math.log10(f_nyquist)

w_p = numpy.logspace(-20,k_nyquist,1000)
[re,im,w] = control.nyquist(g,w_p)

from matplotlib import pyplot as plt

plt.grid(True)
plt.xlabel('Real')
plt.ylabel('Imaginary')
plt.ylim(-10,10)
plt.xlim(-5,12)
plt.show()
```

The obtained plot of the DC motor is as in Figure 7.4. As can be seen in this figure, Nyquist plot does not encircle the -1 point. Moreover, the DC motor has no poles outside the Nyquist path. Hence, it is stable (Phillips et al. 2015).

Figure 7.5 General structure of the open-loop control system.

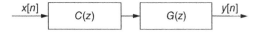

7.2 The Effect of Open-Loop Control on System Performance

The system at hand may not satisfy the desired performance criteria introduced in the previous section. Therefore, a controller should be added to it. We will introduce the open-loop control strategy in this section. Therefore, we will start with explaining what an open-loop controller is. Then, we will analyze the effect of open-loop controller on system performance.

7.2.1 What is Open-Loop Control?

The system at hand may not produce the desired output for a given input. Or the desired performance criteria may not be satisfied by the system alone. Therefore, a controller may be needed which will be placed between the input (control) signal and actual input of the system. For such cases, the easiest solution is designing a controller by open-loop control strategy. The idea here is modifying the control input before feeding it to the system. The new structure will be formed as in Figure 7.5. In this new setup, the output can be obtained as $Y(z) = G(z)C(z)X(z)$ where $C(z)$ is the controller. The aim here is modifying the actual control input signal $x[n]$ by $C(z)$ before feeding it to the system $G(z)$ such that it gives the desired output. The controller $C(z)$ will be designed using the open-loop control strategy. This methodology works best when there is no disturbance affecting the system or the system model at hand is fairly robust to external disturbances.

We can benefit from MicroPython control systems library to form an open-loop controller and connect it to the actual system. We can use the function `series()` within the library for this purpose. Let us provide such an example on the DC motor considered thus far.

Example 7.9 *Applying open-loop control to the DC motor*
Let us take transfer function of the DC motor in Example 7.1 and apply an open-loop controller with the transfer function $C(z) = 0.1$. We can use the Python code in Listing 7.7 on PC to obtain the equivalent transfer function after applying the controller.

Listing 7.7: Applying open-loop control to the DC motor via MicroPython control systems library.

```
import mpcontrolPC

#System
```

```
Ts = 0.0005
num1 = [0.00943, 0.01886, 0.00943]
den1 = [1, -1.8188, 0.8224]

# Transfer function of the system
g = mpcontrolPC.tf(num1,den1,Ts)

#Open-loop controller, K=0.1
num2 = [0.1]
den2 = [1]

c = mpcontrolPC.tf(num2,den2,Ts)

# Transfer function of the open-loop controlled system
gol = mpcontrolPC.series(g,c)
```

As we execute the code in Listing 7.7, we obtain the overall transfer function of the DC motor with open-loop controller as

$$G_{ol}(z) = \frac{0.0009z^2 + 0.0019z + 0.0009}{z^2 - 1.8188z + 0.8224} \tag{7.4}$$

7.2.2 Improving the System Performance by Open-Loop Control

We can analyze how an open-loop controller affects the overall system performance. Therefore, we pick a predesigned controller in Example 7.9. We first analyze the improvement of the overall system in time domain. Then, we observe the improvement in frequency domain. The emphasis here is showing how open-loop control affects the performance criteria in both time and frequency domains. Let us start with time-domain analysis.

Example 7.10 *Improving the performance of DC motor by open-loop control*
Let us focus on the modified transfer function in Eq. (7.4) and see how the open-loop controller affects step response of the overall system. We can obtain the unit step response of the DC motor after applying the open-loop controller. To do so, we can benefit from the Python code in Listing 7.8 on PC.

Listing 7.8: Obtaining unit-step response of the DC motor after applying the open-loop controller to it.

```
import mpcontrol
import mpcontrolPC

#System
Ts = 0.0005
num1 = [0.00943, 0.01886, 0.00943]
den1 = [1, -1.8188, 0.8224]

# Transfer function of the system
g = mpcontrolPC.tf(num1,den1,Ts)

num2 = [0.1]
```

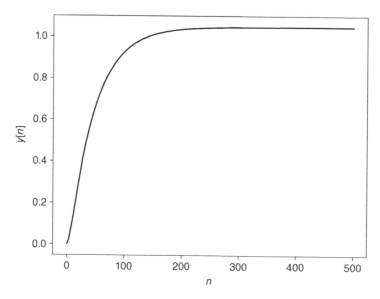

Figure 7.6 Unit step response of the DC motor with open-loop controller applied to it.

```
den2 = [1]

# Transfer function of the open-loop controller, K=0.1
c = mpcontrolPC.tf(num2,den2,Ts)

# Transfer function of the open-loop controlled system
gol = mpcontrolPC.series(g,c)

#Step input signal
N = 500
x = mpcontrol.step_signal(N, 1)

#Calculate the output signal
y = mpcontrol.lsim(gol, x)

gol.stepinfo()

ss_error = gol.ss_error('Step')
```

As we execute the code in Listing 7.8, we obtain unit-step response of the DC motor with open-loop controller applied to it as in Figure 7.6.

We can add the code line `gol.stepinfo()` to the Python code in Listing 7.8 to obtain transient response measures of the DC motor with the open-loop controller applied to it. As we execute this code, we obtain T_{rt}, M_p, T_{pt}, T_{st}, and e_{ss} values as in Table 7.1. As can be seen in this table, the e_{ss} value decreased when an open-loop controller is used. No change has been observed in T_{rt}, M_p, T_{pt}, and T_{st} values.

Table 7.1 The effect of open-loop control in time domain.

System	T_{rt} (sec)	M_p (%)	T_{pt} (sec)	T_{st} (sec)	e_{ss}
no controller	0.0485	0	0.4345	0.088	-9.4778
open-loop control	0.0485	0	0.4345	0.088	-0.0478

We can also analyze the improvement of the overall system in frequency domain when the open-loop controller is applied to the DC motor. To do so, we obtain Bode plot and provide comparison of frequency domain parameters before and after the open-loop controller is applied to the DC motor.

Example 7.11 *Bode plot of the DC motor after applying the open-loop controller to it*

We can use the Python code in Listing 7.9 on PC to obtain Bode plot of the DC motor with open-loop controller applied to it.

Listing 7.9: Obtaining Bode plot of the DC motor after applying the open-loop controller to it.

```
import mpcontrolPC

#System
Ts = 0.0005
num1 = [0.00943, 0.01886, 0.00943]
den1 = [1, -1.8188, 0.8224]

# Transfer function of the system
g = mpcontrolPC.tf(num1,den1,Ts)

num2 = [0.1]
den2 = [1]

# Transfer function of the open-loop controller, K=0.1
c = mpcontrolPC.tf(num2,den2,Ts)

# Transfer function of the open-loop controlled system
gol = mpcontrolPC.series(g,c)

#Bode plot of the system
[mag, phase, freq] =  mpcontrolPC.bode(gol,1000)
```

As we execute the code in Listing 7.9, we obtain Bode plot of the system both in magnitude and phase terms for 1000 elements. We use the `matplotlib` library to plot the magnitude and phase terms in Bode plot in Figures 7.7(a) and 7.7(b), respectively.

Although Figure 7.7 is helpful in understanding the effect of open-loop controller to the DC motor, it is better to obtain frequency domain measures to quantify this effect. Therefore, we can add the function `margin()` to the Python code in Listing 7.9. As the modified code is executed, we obtain GM, PM, and

Table 7.2 The effect of open-loop control in frequency domain.

System	GM (dB)	PM (degree)	ω_{cg} (rad/sec)	ω_{cp} (rad/sec)
no controller	313.071	53.094	6283.190	338.802
open-loop control	313.071	160.428	6283.190	14.229

other related values as in Table 7.2. As can be seen in this table, when we add the open-loop controller with gain 0.1 to the DC motor, the magnitude plot goes down because $20 \log(K) = 20 \log(0.1)$ is approximately -20 dB. This causes a change in ω_{cp}. As a result, the PM value is increased when the open-loop controller is added to the system.

7.3 The Effect of Closed-Loop Control on System Performance

The output obtained in open-loop control is neither measured nor taken into account during operation. This may be undesirable for some cases. Therefore, closed-loop control methodology has emerged. We consider the closed-loop control action in this section. We also analyze the effect of this operation on system performance.

7.3.1 What is Closed-Loop Control?

In closed-loop control, the system output is measured and fed back to be compared with the input (reference) signal to generate the desired control action. Thus, output of the system has an effect on the control action in closed-loop control. This is the main difference between open- and closed-loop control. We provide schematic representation of the closed-loop control in Figure 7.8.

As can be seen in Figure 7.8, output of the open-loop control action, $y[n]$, is fed back to input by another system (sensor most of the times) with the transfer function $H(z)$. This measured signal is fed back to input such that an error signal is formed as $e[n]$. Hence, the closed-loop is formed. Then, $e[n]$ is fed to the controller, $C(z)$, to generate the desired signal to be fed to the system $G(z)$. Based on these definitions, we will have

$$E(z) = X(z) - Y(z)H(z) \tag{7.5}$$

$$Y(z) = E(z)C(z)G(z) \tag{7.6}$$

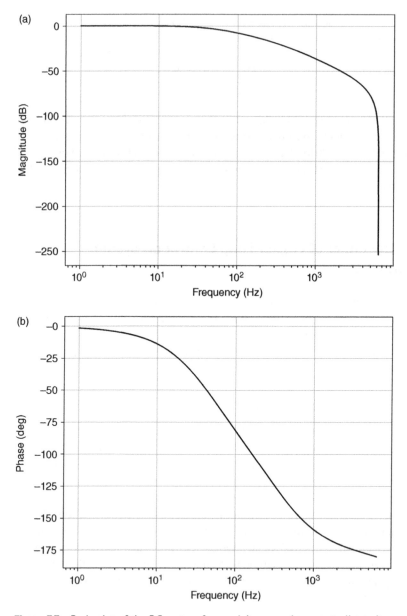

Figure 7.7 Bode plot of the DC motor after applying open-loop controller to it.

Figure 7.8 General structure of the closed-loop control system.

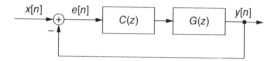

Hence, the overall transfer function of the system can be obtained as

$$\frac{Y(z)}{X(z)} = \frac{C(z)G(z)}{1 + H(z)C(z)G(z)} \tag{7.7}$$

It is possible to operate the DC motor at any speed using open-loop control. However, we cannot obtain accuracy in speed. Or the speed may not be constant for load variations on the motor. Therefore, it may be desirable to feed the output signal back to the system and compare the actual output with the desired one. This way, we can reduce the error or reach the desired value in a shorter time period. We can form the closed-loop controller for this purpose. We can benefit from MicroPython control systems library to construct the closed-loop control action. Let us see how this can be done on the DC motor.

Example 7.12 *Closed-loop control of the DC motor*
Let us take transfer function of the DC motor in Example 7.1 and apply a closed-loop controller to it. We can use the encoder on our DC motor to measure speed of the motor and use this information to reach the desired system performance. To do so, we should first form the closed-loop control setup as in Figure 7.8. Then, we should obtain the transfer function of our encoder as the feedback element. We can take transfer function of the encoder as $H(z) = 1$ without loss of generality. Then, we should design the controller $C(z)$. We will deal with this design procedure in Chapter 8. For now, we can take $C(z) = 1.5$ as the predesigned controller.

We can use the Python code in Listing 7.10 on PC to obtain the equivalent transfer function after applying the controller.

Listing 7.10: Obtaining the closed-loop transfer function of the DC motor.

```
import mpcontrolPC

#System
Ts = 0.0005
num1 = [0.00943, 0.01886, 0.00943]
den1 = [1, -1.8188, 0.8224]

#Transfer function of the system
g = mpcontrolPC.tf(num1,den1,Ts)

#Transfer function of the controller, K=1.5
num2 = [1.5]
den2 = [1]

c = mpcontrolPC.tf(num2,den2,Ts)
```

```
#Transfer function of the feedback system
num3 = [1]
den3 = [1]

h = mpcontrolPC.tf(num3,den3,Ts)

# Transfer function of the closed-loop controlled system
gol = mpcontrolPC.series(g,c)
gcl = mpcontrolPC.feedback(gol,h)
```

As we execute the code in Listing 7.10, we obtain the overall transfer function of the DC motor with closed-loop controller as

$$G_{cl}(z) = \frac{0.0139z^2 + 0.0279z + 0.0139}{z^2 - 1.7655z + 0.8249} \tag{7.8}$$

7.3.2 Improving the System Performance by Closed-Loop Control

Although we will introduce methods to design a closed-loop controller in the next chapter, we will analyze how such a controller affects the overall system performance here. Therefore, we picked a predesigned controller in Example 7.12. In this section, we will first analyze the improvement of the overall system in time domain. Then, we will observe the improvement in frequency domain. The emphasis here is showing how closed-loop control affects the performance criteria both in time and frequency domains. Let us start with the time-domain analysis.

Example 7.13 *Improving the performance of DC motor by closed-loop control*

Let us focus on the modified transfer function in Eq. (7.8) and see how the closed-loop controller affects step response of the overall system. We can obtain unit step response of the DC motor after applying the closed-loop controller. To do so, we can benefit from the Python code in Listing 7.11 on PC.

Listing 7.11: Obtaining unit step response of the DC motor after applying closed-loop controller to it.

```
import mpcontrol
import mpcontrolPC

#System
Ts = 0.0005
num1 = [0.00943, 0.01886, 0.00943]
den1 = [1, -1.8188, 0.8224]

#Transfer function of the system
g = mpcontrolPC.tf(num1,den1,Ts)

#Transfer function of the controller, K=1.5
num2 = [1.5]
```

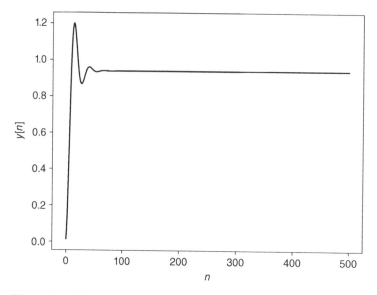

Figure 7.9 Unit step response of the DC motor with closed-loop controller applied to it.

```
den2 = [1]

c = mpcontrolPC.tf(num2,den2,Ts)

#Transfer function of the feedback system
num3 = [1]
den3 = [1]

h = mpcontrolPC.tf(num3,den3,Ts)

# Transfer function of the closed-loop controlled system
gol = mpcontrolPC.series(g,c)
gcl = mpcontrolPC.feedback(gol,h)

#Step input signal
N = 500
x = mpcontrol.step_signal(N, 1)

#Calculate the output signal
y = mpcontrol.lsim(gcl, x)
```

As we execute the code in Listing 7.11, we obtain unit-step response of the DC motor with closed-loop controller applied to it as in Figure 7.9.

We can also add the code lines `gcl.stepinfo()` and `gcl.ss_error()` to the Python code in Listing 7.11 to obtain transient response measures of the DC motor with the closed-loop controller applied to it. As we execute this code, we obtain $T_{rt}, M_p, T_{pt}, T_{st}$, and e_{ss} values as in Table 7.3. As can be seen in this table, T_{rt},

Table 7.3 The effect of closed-loop control in time domain.

System	T_{rt} (sec)	M_p (%)	T_{pt} (sec)	T_{st} (sec)	e_{ss}
no controller	0.0485	0	0.4345	0.0880	-9.4778
open-loop control	0.0485	0	0.4345	0.0880	-0.0478
closed-loop control	0.0025	27.6926	0.0065	0.0205	0.0598

T_{pt}, T_{st}, and e_{ss} values decreased when a closed-loop controller is used. However the M_p value increased.

We can also analyze the improvement of the overall system in frequency domain when the closed-loop controller is applied to the DC motor. To do so, we obtain Bode plot and provide comparison of frequency domain measures before and after the closed-loop controller is applied to the DC motor.

Example 7.14 *Bode plot of the DC motor after applying the closed-loop controller to it*

We can use the Python code in Listing 7.12 on PC to obtain Bode plot of the DC motor with closed-loop controller applied to it.

Listing 7.12: Obtaining Bode plot of the DC motor after applying the closed-loop controller to it.

```
import mpcontrolPC

#System
Ts = 0.0005
num1 = [0.00943, 0.01886, 0.00943]
den1 = [1, -1.8188, 0.8224]

#Transfer function of the system
g = mpcontrolPC.tf(num1,den1,Ts)

#Transfer function of the controller, K=1.5
num2 = [1.5]
den2 = [1]

c = mpcontrolPC.tf(num2,den2,Ts)

#Transfer function of the feedback system
num3 = [1]
den3 = [1]

h = mpcontrolPC.tf(num3,den3,Ts)

# Transfer function of the closed-loop controlled system
gol = mpcontrolPC.series(g,c)
gcl = mpcontrolPC.feedback(gol,h)
```

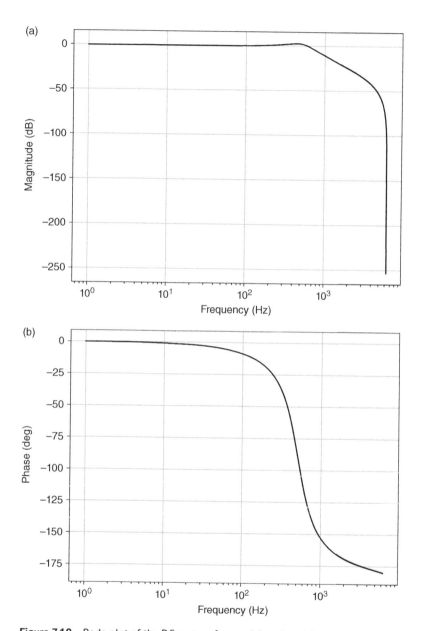

Figure 7.10 Bode plot of the DC motor after applying closed-loop controller to it.

Table 7.4 The effect of closed-loop control in frequency domain.

System	GM (dB)	PM (degree)	ω_{cg} (rad/sec)	ω_{cp} (rad/sec)
no controller	313.071	53.094	6283.190	338.802
open-loop control	313.071	160.428	6283.190	14.229
closed-loop control	313.071	69.013	6283.190	590.175

```
#Bode plot of the system
[mag, phase, freq] = mpcontrolPC.bode(gcl,1000)

#GM, PM, Crossover frequency values of the system
mpcontrolPC.margin(gcl)
```

As we execute the code in Listing 7.12, we obtain Bode plot of the system both in magnitude and phase terms for 1000 elements. We can use the `matplotlib` library to plot the magnitude and phase terms in Bode plot in Figures 7.10(a) and 7.10(b), respectively.

Although Figure 7.10 is helpful in understanding the effect of closed-loop controller to the DC motor, it is better to obtain frequency domain measures to quantify this effect. Therefore, we can add the function `margin()` to the Python code in Listing 7.12. As the code is executed, we obtain GM, PM, and other related values as in Table 7.4.

As can be seen in Table 7.4, when we add the closed-loop controller with gain 1.5 to the DC motor, the magnitude plot changes and as a result there is a change in ω_{cp}. This results in PM as 69.013°. Such a PM is generally selected as starting point for the controller design in many applications since it results in fast settling time, T_{st}.

7.4 Application: Adding Open-Loop Digital Controller to the DC Motor

The aim of this application is to show how a digital controller realized on the STM32 microcontroller can improve DC motor characteristics. We provide the system setup for this application in Figure 7.11. Here, the "System Controller" block inside the microcontroller contains functions to realize the digital controller. We will provide the detailed description of these functions while explaining C and Python codes of the application.

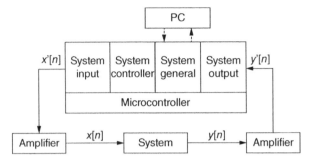

Figure 7.11 System setup for the application.

7.4.1 Hardware Setup

The hardware setup in this application is the same as in Section 4.7. The only difference here is that, there will be a controller block between the microcontroller output and DC motor input. To note here, the controller block is realized in software not in hardware.

7.4.2 Procedure

In this application, we pick the open-loop control strategy by selecting the controller transfer function as $C(z) = 0.0923$. We apply a constant amplitude input to the DC motor as 60 rpm. Then, we observe speed of the DC motor from its quadrature encoder and send the observed data to PC in real-time for one second. Next, we use transfer function of the DC motor obtained in Chapter 6. Hence, we simulate the system with open-loop controller by applying the same input signal. Finally, we compare the simulated and actual output signals as plots and transient response measures.

7.4.3 C Code for the System

We can form a new Mbed Studio project in C language to apply the open-loop controller to the DC motor. The software setup here will be the same as in Section 3.4.3.

As we check the `system_controller.cpp` file, there are two functions of interest. These are explained briefly as follows: The `controller_tf_init` `(float *num, float *den, int nbr_num_coeffs, int nbr_den_` `coeffs, int select)` function is used to create discrete-time controller and select between open-loop or closed-loop control by the `select` input. If `select` is 0, then open-loop controller is realized. If `select` is 1, then closed-loop controller is realized. The `obtain_controller_output(float input,` `float output, int cnt_signal)` function is used to generate controller output for a given input, and output if feedback is used.

We should add the function `controller_tf_init` to the `main` function to create and add a controller to the system. For this application, we set `select` as 0 since we will apply the open-loop control strategy. We choose a controller with constant gain 0.0923. This represents the relation between motor voltage and speed. This value is obtained from the DC motor datasheet as 12V/130 rpm. Here, the controller input will be the desired speed value. Hence, it will be entered as constant amplitude in the terminal interface. The modified main file will be as in Listing 7.13.

Listing 7.13: Modified C code for the open-loop controller.

```
#include "mbed.h"
#include "system_general.h"
#include "system_input.h"
#include "system_output.h"
#include "system_controller.h"

Serial pc(USBTX, USBRX, 115200);
RawSerial device(PD_5, PD_6, 921600);
Ticker sampling_timer;
InterruptIn button(BUTTON1);
Timer debounce;
DigitalOut led(LED1);
PwmOut pwm1(PE_9);
PwmOut pwm2(PE_11);
DigitalOut pwm_enable(PF_15);
InterruptIn signal1(PA_0);
InterruptIn signal2(PB_9);
Timer encoder_timer;
AnalogIn ADCin(PA_3);
AnalogOut DACout(PA_4);

int main()
{
    float num = 0.0923;
    float den = 1;
    user_button_init();
    system_input_init(1, 12/3.3);
    system_output_init(1, 3591.84);
    send_data_RT(2);
    sampling_timer_init();
    controller_tf_init(&num, &den, 1, 1, 0);
    menu_init();
    while (true) {
        system_loop();
    }
}
```

In order to compare the real system output and simulation results, the reader should run the Python code in Listing 7.14. Then, the user button on the STM32 board should be pressed. Afterward, the DC motor starts running and data transfer to PC begins.

Listing 7.14: Python code to compare the real system output and simulation results.

```
import mpcontrol
import mpcontrolPC

N = 2000
signal1, signal2 = mpcontrolPC.get_data_from_MC(N, 'com10', 1)
mpcontrolPC.save_to_file(signal1, file_name='input.dat')
mpcontrolPC.save_to_file(signal2, file_name='output.dat')

signal3 = mpcontrolPC.ma_filter(signal2, 15)
mpcontrolPC.save_to_file(signal3, file_name='output_filtered.dat')

Ts = 0.0005
num = [0.00943, 0.01886, 0.00943]
den = [1, -1.8188, 0.8224]
g = mpcontrolPC.tf(num, den, Ts)

Kp = 0.0923
num_cc = [Kp]
den_cc = [1.0]
tf_Pcontroller = mpcontrolPC.tf(num_cc, den_cc, Ts)
tf_forward = mpcontrolPC.series(tf_Pcontroller, g)

y = mpcontrolPC.compare_stepinfo(signal3, tf_forward, 60)

mpcontrolPC.save_to_file(y, file_name='output_simulation.dat')
```

In Listing 7.14, we should set `get_data_from_MC(NoElements, com_port, selection)` for real-time data transfer for two signals. Then, the function `save_to_file()` is used to save the received data to files. Afterward, we construct digital transfer function of the DC motor. Then, we obtain time domain performance criteria for both simulation and actual output data using the function `compare_stepinfo(Real_Time_Output, Transfer_Function, Step_Gain)`.

7.4.4 Python Code for the System

We can start the application by embedding the Python code on the STM32 microcontroller. To do so, we should replace content of the `main.py` file by the Python code in Listing 7.15 by following the steps explained in Section 3.2.2. We should also embed the `system_general.py`, `system_input.py`, `system_output.py`, `system_controller.py`, and `mpcontrol.py` files to flash. To note here, these files are provided in the accompanying book website. As a reminder, Python code to be used on the PC side is the same as in Section 7.4.3.

Listing 7.15: Python code to be used in open-loop control of the DC motor.

```
import system_general
import system_input
import system_output
import system_controller
```

```
def main():
        system_general.button_init()
        system_general.PWM_frequency_select(0)
        system_general.sampling_timer_init(0)
        system_general.input_select(input_type='Constant Amplitude',
            amplitude=40)
        system_controller.controller_init([0.0923], [1], 0.0005, 0)
        system_general.motor_direction_select(1)
        system_input.system_input_init(2, 3.63636363636)
        system_output.system_output_init(1, 3591.84)
        system_general.send_data_RT(2)
        while True:
                system_general.system_loop()
main()
```

If we check the `system_controller.py` file, there are two new functions. The function `controller_init(num, den, ts, select)` is used to create the digital controller and select between open- or closed-loop control strategy. If `select` is set to 0, then open-loop control is applied. If `select` is set to 1, then closed-loop control is applied. The `controller_output(input)` function is used to generate the controller output for a given input.

7.4.5 Observing Outputs

When we execute the C code of the application, the resulting signals will be as in Figure 7.12. We can obtain similar results when we execute the Python code for this application as well. As can be seen in Figure 7.12, the actual and simulated outputs are almost the same. Unfortunately, the actual output is noisy. For both cases, the desired speed cannot be obtained at output. This indicates that the open-loop strategy is not sufficient for the DC motor to decrease e_{ss}. Therefore, we will consider the closed-loop control strategy in Chapter 8.

We tabulate the time domain performance criteria for the actual and simulation outputs in Table 7.5. To note here, since the actual output has noise it affects the obtained values. Therefore, the tabulated values may change slightly. As can be seen in the table, the actual and simulated outputs are close to each other.

7.5 Summary

We focused on transfer function-based control systems in this chapter. We introduced definitions for both open- and closed-loop control systems based on transfer functions. We also introduced performance criteria based on transfer functions. As the end of chapter application, we applied open-loop control strategy to the DC motor both in Python and C languages. We observed the actual and simulated system output. This application forms the backbone for all digital open- and

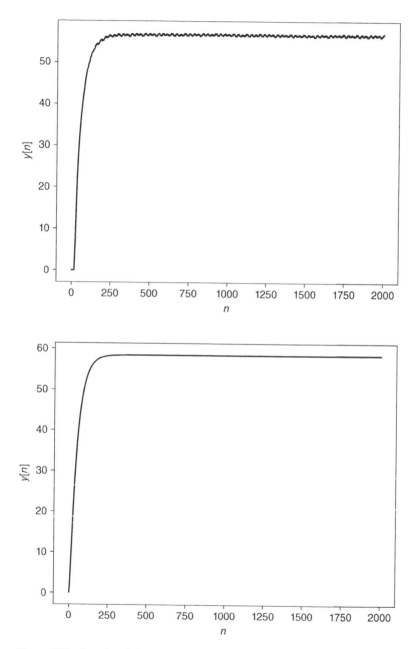

Figure 7.12 Actual and simulated output signals obtained from the DC motor.

Table 7.5 Time domain performance criteria for the DC motor.

Measure	Actual	Simulation
Rise Time (T_{rt}), sec	0.0460	0.0485
Settling Time (T_{st}), sec	0.1010	0.0880
Peak Time (T_{pt}), sec	0.4570	0.4345
Overshoot (M_p), %	0	0
Undershoot, %	0	0
Settling min.	51.0395	52.2299
Settling max.	57.1000	58.0259
Peak	57.1000	58.0259

closed-loop controller implementations. Besides, the setup introduced here is not specific to the DC motor. The reader can use it to control his or her system. The part to be added here is the digital controller. We will introduce controller design techniques based on transfer functions in the next chapter.

Problems

7.1 Consider a system with transfer function

$$G(z) = \frac{1}{5z^3 - 3z^2 + 2z + 1}$$

Use Python and MicroPython control systems libraries to
a. find unit step response of the system.
b. calculate ζ and ω_n values of the system.
c. determine T_{rt}, M_p, T_{pt}, and T_{st} of the system.
d. find e_{ss} of the system.

7.2 Consider a continuous-time system with transfer function

$$G(s) = \frac{20(s + 5)}{s(s + 4)(s + 6) + 25}$$

Use Python and MicroPython control systems libraries to
a. obtain the digital representation of the systems using bilinear transformation. Set $T_s = 0.05$ sec for transformation.
b. plot unit step response of the digital system.
c. calculate ζ and ω_n values of the digital system.
d. determine T_{rt}, M_p, T_{pt}, and T_{st} of the digital system.
e. find e_{ss} of the digital system.

Figure 7.13 The continuous-time system in Problem 7.6.

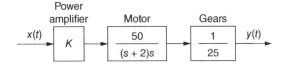

Figure 7.14 The system in Problem 7.8.

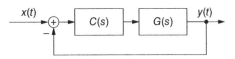

7.3 Repeat Problem 7.2 when $T_s = 0.02$ sec. What has changed?

7.4 Sketch Bode plot of the system given in Problem 7.2. Calculate PM and GM of the system. Comment on stability of the system based on these values.

7.5 Plot Nyquist plot of the system given in Problem 7.2. Based on this plot, determine whether the system is stable or not.

7.6 Consider the continuous-time system in Figure 7.13.
Use Python and MicroPython control systems libraries to
a. obtain the digital transfer function $(Y(z))/(X(z))$ of the system in Figure 7.13 by using the zero-order hold method. Set K = 2.4 and $T_s = 0.1$ sec in operation.
b. plot unit-step response of the digital system.
c. calculate ζ and ω_n values of the digital system.
d. determine T_{rt}, M_p, T_{pt}, and T_{st} values of the digital system.
e. find e_{ss} of the digital system.

7.7 Reconsider the digital system in Problem 7.6 using Python and MicroPython control systems libraries to
a. sketch Bode plot of the system. Calculate PM and GM of the system. Comment on stability of the system based on these values.
b. plot root locus plot of the system.
c. obtain Nyquist plot of the system and determine whether it is stable or not based on this plot.

7.8 Consider the system in Figure 7.14.
where
$$C(s) = \frac{10}{s+10}$$
$$G(s) = \frac{K}{s(s+50)}$$

and $10 \leq K \leq 40$, $T_s = 0.1$ sec.

Use Python and MicroPython control systems libraries to

a. find the digital transfer function $(Y(z)/X(z))$ of the system for K values 10, 20, and 40.
b. plot unit-step response of the systems obtained in part (a).
c. determine T_{rt}, M_p, T_{pt}, T_{st}, and e_{ss} of the systems in part (a). Comment on the results for each K value.

7.9 The feedforward transfer function of a unity feedback closed loop system is

$$C(z)G(z) = \frac{0.3625z}{(z-1)(z-0.4562)}$$

Use Python and MicroPython control systems libraries to

a. plot unit step response of the closed-loop system.
b. calculate ζ and ω_n values of the closed-loop system.
c. determine T_{rt}, M_p, T_{pt}, and T_{st} of the closed-loop system.
d. find e_{ss} of the closed-loop system.

8

Transfer Function Based Controller Design

The linear and time-invariant (LTI) system at hand may not satisfy the required design criteria. To solve this problem, a controller can be added to it as mentioned in Chapter 7. There, we picked predesigned controllers. In this chapter, we will explore the transfer function-based controller design methodology for this purpose. We will benefit from the closed-loop control strategy in the design phase throughout the chapter since it is applicable to a broader class of problems. As a reminder, the closed-loop controller feeds the output signal back to input. Hence, the reference and actual output signals are compared to generate the control signal. We will formalize the transfer function based controller design method this way. As for the controller structure, we will first introduce the proportional integral derivative (PID) controller and its versions as proportional (P) and proportional integral (PI) controllers. We will provide Python based design methods for these. Then, we will introduce the lag–lead controller and its versions as the lag and lead controllers. We will also provide their design methodology in MATLAB. As the end of chapter application, we will design transfer function-based controllers to the actual DC motor and observe their effects on the performance criteria.

8.1 PID Controller Structure

PID is one of the most popular controller design structure used in practical applications. The main reason for this popularity is that PID has well-defined design steps. Hence, the user can follow them to construct a fairly good working controller.

The continuous-time PID controller transfer function is represented as

$$C(s) = C_p(s) + C_i(s) + C_d(s) \tag{8.1}$$

$$C_p(s) = K_p \tag{8.2}$$

Embedded Digital Control with Microcontrollers: Implementation with C and Python,
First Edition. Cem Ünsalan, Duygun E. Barkana, and H. Deniz Gürhan.
© 2021 The Institute of Electrical and Electronics Engineers, Inc. Published 2021 by John Wiley & Sons, Inc.
Companion website: www.wiley.com/go/Unsalan/Embedded_Digital_Control_with_Microcontrollers

$$C_i(s) = \frac{K_i}{s} \tag{8.3}$$

$$C_d(s) = K_d s \tag{8.4}$$

where K_p, K_i, and K_d stand for gain parameters to be set for the design criteria at hand. To note here, one or two of these gains can be set to zero to obtain the PI and P controllers, respectively.

The transfer function in Eq. (8.1) can be converted to digital form by using one of the conversion methods, preferably bilinear transformation, discussed in Section 5.3. Hence, the PID controller can be used in digital domain.

We will introduce the PID controller design strategy in this section. To do so, we will first focus on the P, PI, and PID controllers separately. Then, we will introduce three parameter tuning methods for these controllers.

8.1.1 The P Controller

The aim of the proportional (P) controller is to multiply the error signal by a constant in the closed-loop setup and generate the control signal accordingly. Based on this, the closed-loop control with the P controller will be as in Figure 8.1.

The controller $C_p(z)$ in Figure 8.1 is obtained by converting the transfer function $C_p(s) = K_p$ to digital form by conversion methods introduced in Section 5.3. To note here, K_p in the P controller will be obtained by parameter tuning methods to be introduced in Section 8.1.4.

8.1.2 The PI Controller

We can use the PI controller when the P controller is not sufficient for the desired design criteria. To do so, the PI controller benefits from the past information by adding an integrator path in the controller. This is done by adding a zero at $-(K_i/K_p)$ and pole at origin. The system with this controller will have large T_{rt} and T_{st}. The closed-loop control with the PI controller will be as in Figure 8.2.

The controller $C_p(z)$ in Figure 8.2 is the same as in the P controller. In this figure, the controller $C_i(z)$ is obtained by converting the transfer function $C_i(s) = \frac{K_i}{s}$ to digital form by conversion methods introduced in Section 5.3. To note here, K_p

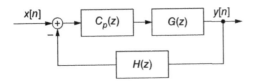

Figure 8.1 Closed-loop control with the P controller.

Figure 8.2 Closed-loop control
with the PI controller.

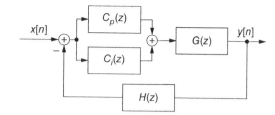

and K_i in the PI controller will be obtained by parameter tuning methods to be introduced in Section 8.1.4.

8.1.3 The PID Controller

The PI controller can be extended further by adding a derivative term. This leads to the PID controller which merges both previous input values and future predictions in the control action. The derivative term improves the controller by predicting future input values. However, it may also have a negative effect if the signal to be processed is noisy. For such cases, the derivative term may enhance the noise in the signal. As a result, the PID controller may not work as expected or may not work at all in practice. Therefore, this structure should be used with precaution. The closed-loop control setup with the PID controller will be as in Figure 8.3.

The controllers $C_p(z)$ and $C_i(z)$ in Figure 8.2 are the same as in the PI controller. In this figure, the controller $C_d(z)$ is obtained by converting the transfer function $C_d(s) = K_d s$ to digital form by one of the conversion methods introduced in Section 5.3. To note here, K_p, K_i, and K_d in the PID controller will be obtained by parameter tuning methods to be introduced in Section 8.1.4.

8.1.4 Parameter Tuning Methods

The PID controller has three gain parameters to be set as introduced in the previous section. For this purpose, there are three well-known methods in literature as Ziegler–Nichols, Cohen–Coon, and Chien–Hrones–Reswick. We will introduce them in this section. Let us start with the Ziegler–Nichols method.

Figure 8.3 Closed-loop
control with the PID
controller.

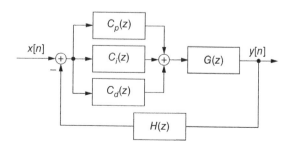

8.1.4.1 The Ziegler–Nichols Method

To set the PID gain parameters, the Ziegler–Nichols method uses three auxiliary parameters obtained from response of the system to be controlled (Ziegler and Nichols 1942). These are K, L, and a. The K, L, and τ (to be used in obtaining a) parameters are calculated using the first-order dead-time model $G(s) = Ke^{-Ls}/(\tau s + 1)$. Here, K is the system gain, L is the system dead-time, and τ is the time constant. The K parameter is calculated as $K = \frac{y(\infty)-y(0)}{x(\infty)-x(0)}$ where $y(\infty)$ and $y(0)$ are the final and initial values of the system response, respectively. Likewise, $x(\infty)$ and $x(0)$ are the final and initial values of the system input, respectively. The L and τ parameters are related to step response of the system. We can estimate τ by multiplying the time it takes the system to reach 28.3% of its step response by 1.5. We can find L by subtracting τ from the time it takes the system to reach 63.2% of its step response. Finally, the a parameter is obtained as $a = K\frac{L}{\tau}$.

As we obtain K, L, and a, we can calculate the PID gain parameters as tabulated in Table 8.1. The reader can pick the parameter set from this table for the selected controller type as P, PI, or PID.

8.1.4.2 The Cohen–Coon Method

The Cohen–Coon method is similar to the Ziegler–Nichols method with one difference (Cohen and Coon 1953). Besides the previously introduced parameters a and L, this method has an extra parameter $R = \frac{L}{\tau}$ for operation. Based on these, we should first calculate the auxiliary parameters, given in Table 8.2, to calculate the actual PID gain parameters. Then, we can calculate the PID gain parameters as tabulated in Table 8.3. The reader can pick the parameter set from this table for the selected controller type as P, PI, or PID.

8.1.4.3 The Chien–Hrones–Reswick Method

The third and final parameter tuning method is offered by Chien et al. (1952). Auxiliary parameters to be used in this method can be calculated as in Table 8.4. Then, we can calculate the PID gain parameters as tabulated in Table 8.5. The reader can pick the parameter set from this table for the selected controller type as P, PI, or PID.

Table 8.1 PID gain parameters according to the Ziegler–Nichols method.

Controller	K_p	K_i	K_d
P	$1.0/a$	—	—
PI	$0.9/a$	$K_p/(3L)$	—
PID	$1.2/a$	$K_p/(2L)$	$K_p L/2$

Table 8.2 Auxiliary parameters for the Cohen–Coon method.

Controller	T_i	T_d
PI	$L((30 + 3R)/(9 + 20R))$	—
PID	$L((32 + 6R)/(13 + 8R))$	$4L/(11 + 2R)$

Table 8.3 PID gain parameters according to the Cohen–Coon method.

Controller	K_p	K_i	K_d
P	$(1/a)(1 + (R/3))$	—	—
PI	$(1/a)((9/10) + (R/12))$	K_p/T_i	—
PID	$(1/a)((4/3) + (R/4))$	K_p/T_i	$K_p T_d$

Table 8.4 Auxiliary parameters for the Chien–Hrones–Reswick method.

Controller	T_i	T_d
PI	$2.3L$	—
PID	$2L$	$0.42L$

Table 8.5 PID gain parameters according to the Chien–Hrones–Reswick method.

Controller	K_p	K_i	K_d
P	$0.7/a$	—	—
PI	$0.7/a$	K_p/T_i	—
PID	$1.2/a$	K_p/T_i	$K_p T_d$

8.2 PID Controller Design in Python

The PID controller and parameter tuning methods introduced in the previous section are available as functions in the MicroPython control systems library. We explain their usage in this section. Since the parameters should be set first to implement the P, PI, or PID controller, we start with parameter tuning. Then, we design the controller and evaluate its effect on the closed-loop characteristics of the overall system at hand.

8.2.1 Parameter Tuning

The function `pid_tune(tf1, method, pid)` in the MicroPython control systems library can be used to obtain the PID gain parameters for the selected parameter tuning method and controller type. In this function, `tf1` stands for the continuous-time system to be controlled. `method` stands for the parameter tuning method as `'ZieglerNichols'`, `'CohenCoon'`, or `'CHR'` for the Ziegler–Nichols, Cohen–Coon, and Chien–Hrones–Reswick methods, respectively. The parameter `pid` is used to select the controller type as `'P'`, `'PI'`, or `'PID'`.

8.2.2 Controller Design

We can observe the effects of designed PID controller for a specific system at hand. To do so, we pick the continuous-time transfer function of the DC motor obtained in Chapter 6. We apply P, PI, and PID controllers to it next.

8.2.2.1 P Controller

Let us start with obtaining the P controller gain parameter for the DC motor described by Eq. (6.17) in continuous-time. We can use the Python code in Listing 8.1 for this purpose. Here, we use the Ziegler–Nichols method to obtain the K_p parameter of the controller. Then, we form the controller in continuous-time and convert it to digital form using bilinear transformation.

Listing 8.1: Obtaining the P controller for the DC motor.

```
import mpcontrolPC

#Continous-time System
num = [165702.47]
den = [1, 390.2, 15701]

gs = mpcontrolPC.tf(num, den)

[Kp, Ki, Kd] = mpcontrolPC.pid_tune(gs, 'ZieglerNichols', 'P')

#Analog P Controller
num2 = [Kp]
den2 = [1]
cs = mpcontrolPC.tf(num2, den2)

#Digital P Controller
Ts = 0.0005
cz = mpcontrolPC.c2d(cs, Ts)
```

As the Python code in Listing 8.1 is executed, we obtain the P controller gain parameter as $K_p = 0.7294$. Based on the used functions, the P controller in digital form becomes $C(z) = 0.7294$.

We can obtain unit-step response of the formed closed-loop system using the Python code in Listing 8.2. As we execute this code, we plot the obtained response in Figure 8.4.

Listing 8.2: Unit step response of the closed-loop system formed by the P controller for DC motor.

```
import mpcontrol
import mpcontrolPC

#System
Ts = 0.0005
num1 = [0.00943, 0.01886, 0.00943]
den1 = [1, -1.8188, 0.8224]

#Transfer function of the system
g = mpcontrolPC.tf(num1,den1,Ts)
print(g)

#Transfer function of the controller, Kp
num2 = [0.729409]
den2 = [1]

c = mpcontrolPC.tf(num2,den2,Ts)

#Transfer function of the feedback system
num3 = [1]
den3 = [1]

h = mpcontrolPC.tf(num3,den3,Ts)

# Transfer function of the closed-loop controlled system
gol = mpcontrolPC.series(g,c)
gcl = mpcontrolPC.feedback(gol,h)

#Step input signal
N = 500
x = mpcontrol.step_signal(N, 1)

#Calculate the output signal
y = mpcontrol.lsim(gcl, x)
```

We can calculate the closed-loop system time domain performance criteria using the Python code in Listing 8.3. We will analyze these in a comparative way in Section 8.2.3.

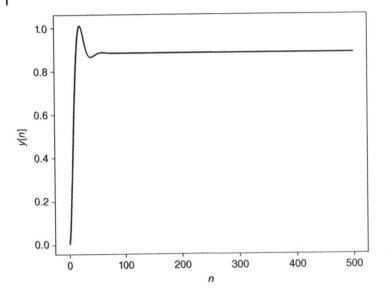

Figure 8.4 Unit step response of the closed-loop system formed by the P controller.

Listing 8.3: Closed-loop system time domain performance criteria for the P controlled DC motor.

```
import mpcontrolPC

#System
Ts = 0.0005
num1 = [0.00943, 0.01886, 0.00943]
den1 = [1, -1.8188, 0.8224]

#Transfer function of the system
g = mpcontrolPC.tf(num1,den1,Ts)
print(g)

#Transfer function of the controller, Kp
num2 = [0.729409]
den2 = [1]

c = mpcontrolPC.tf(num2,den2,Ts)

#Transfer function of the feedback system
num3 = [1]
den3 = [1]

h = mpcontrolPC.tf(num3,den3,Ts)

# Transfer function of the closed-loop controlled system
gol = mpcontrolPC.series(g,c)
gcl = mpcontrolPC.feedback(gol,h)

gcl.damp()

gcl.stepinfo()

ss_error = gcl.ss_error('Step')
```

We can also analyze the closed-loop system formed by the P controller in frequency domain. To do so, we can use the Python code in Listing 8.4 on PC to obtain Bode plot of the DC motor with P controller applied to it. As we execute the Python code in Listing 8.4, we obtain Bode plot of the system both in magnitude and phase terms for 1000 elements as in Figures 8.5(a) and 8.5(b) respectively.

Listing 8.4: Bode plot of the closed-loop controlled DC motor with P controller.

```
import mpcontrolPC

#System
Ts = 0.0005
num1 = [0.00943, 0.01886, 0.00943]
den1 = [1, -1.8188, 0.8224]

#Transfer function of the system
g = mpcontrolPC.tf(num1,den1,Ts)

#Transfer function of the controller, Kp
num2 = [0.729409]
den2 = [1]

c = mpcontrolPC.tf(num2,den2,Ts)

#Transfer function of the feedback system
num3 = [1]
den3 = [1]

h = mpcontrolPC.tf(num3,den3,Ts)

# Transfer function of the closed-loop controlled system
gol = mpcontrolPC.series(g,c)
gcl = mpcontrolPC.feedback(gol,h)

#Bode plot of the system
[mag, phase, freq] = mpcontrolPC.bode(gcl,1000)

#GM, PM, Crossover frequency values of the system
mpcontrolPC.margin(gcl)
```

Although Figure 8.5 is helpful in understanding the closed-loop system formed by the P controller to DC motor, it is better to obtain frequency domain measures. Therefore, we can add the function margin() to the Python code in Listing 8.4. We will analyze the frequency domain performance criteria in a comparative way in Section 8.2.3.

8.2.2.2 PI Controller

Designing the PI controller is the same as in the P controller. We can obtain the controller gain parameters using the Python code in Listing 8.5 for this purpose. Here, we use the Ziegler–Nichols method to obtain the K_p and K_i parameters of the controller. Then, we form the controller in continuous-time and convert it to digital form using bilinear transformation.

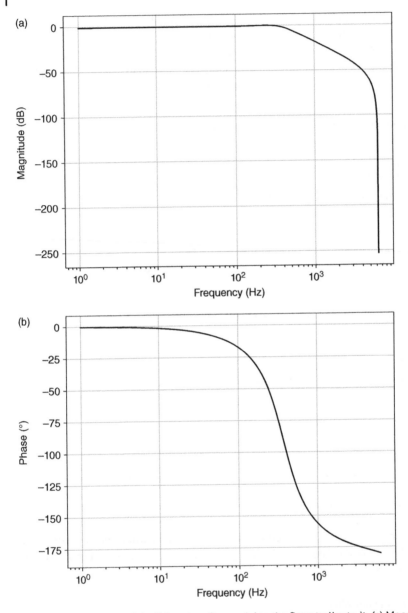

Figure 8.5 Bode plot of the DC motor after applying the P controller to it. (a) Magnitude plot. (b) Phase plot.

Listing 8.5: Obtaining PI controller for the DC motor.

```
import mpcontrolPC

#Continous-time System
num = [165702.47]
den = [1, 390.2, 15701]

gs = mpcontrolPC.tf(num, den)

[Kp, Ki, Kd] = mpcontrolPC.pid_tune(gs, 'ZieglerNichols', 'PI')

#Analog PI Controller
num2 = [Kp]
den2 = [1]

cps = mpcontrolPC.tf(num2, den2)

num3 = [Ki]
den3 = [1, 0]

cis = mpcontrolPC.tf(num3, den3)

#Digital PI Controller
Ts = 0.0005

cpz = mpcontrolPC.c2d(cps, Ts)
ciz = mpcontrolPC.c2d(cis, Ts)
cz = mpcontrolPC.parallel(cpz, ciz)
```

As the Python code in Listing 8.5 is executed, we obtain the PI controller gain parameters as $K_p = 0.6565$ and $K_i = 76.0130$. Based on these values and bilinear transformation, the PI controller in digital form becomes

$$C(z) = \frac{0.6755z - 0.6375}{z - 1} \tag{8.5}$$

We can obtain unit-step response of the closed-loop system using the Python code in Listing 8.2. To do so, we should modify transfer function of the controller as num2 = [0.6755, -0.6375] and den2 = [1, -1]. We plot the obtained response in Figure 8.6.

As in the P controller, we can calculate the closed-loop system time domain performance criteria using the Python code in Listing 8.3. To do so, we should modify transfer function of the controller as num2 = [0.6755, -0.6375] and den2 = [1, -1]. We will analyze the time domain performance criteria in a comparative way in Section 8.2.3.

In order to analyze the closed-loop system formed by the PI controller in frequency domain, we should modify transfer function of the controller in Listing 8.4 as num2 = [0.6755, -0.6375] and den2 = [1, -1]. As we execute this code, we obtain Bode plot of the system both in magnitude and phase terms for 1000 elements as in Figures 8.7(a) and 8.7(b), respectively. We can add the function margin() to the Python code in Listing 8.4 and obtain frequency domain performance measures. We will analyze them in a comparative way in Section 8.2.3.

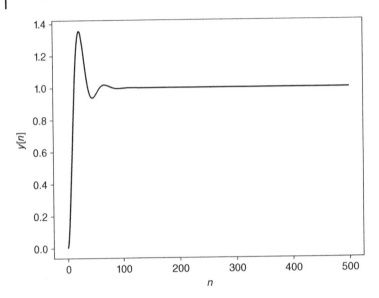

Figure 8.6 Unit step response of the closed-loop system formed by the PI controller.

8.2.2.3 PID Controller

Designing the PID controller is the final step in this section. To do so, we can obtain PID controller gain parameters using the Python code in Listing 8.6. Here, we use the Ziegler–Nichols method to obtain the K_p, K_i, and K_d parameters of the controller. Then, we form the controller in continuous-time and convert it to digital form using bilinear transformation.

Listing 8.6: Obtaining PID controller for the DC motor in Python.

```
import mpcontrolPC

#Continous-time System
num = [165702.47]
den = [1, 390.2, 15701]

gs = mpcontrolPC.tf(num, den)

[Kp, Ki, Kd] = mpcontrolPC.pid_tune(gs, 'ZieglerNichols', 'PID')

#Analog PID Controller
num2 = [Kp]
den2 = [1]

cps = mpcontrolPC.tf(num2, den2)

num3 = [Ki]
den3 = [1, 0]

cis = mpcontrolPC.tf(num3, den3)
```

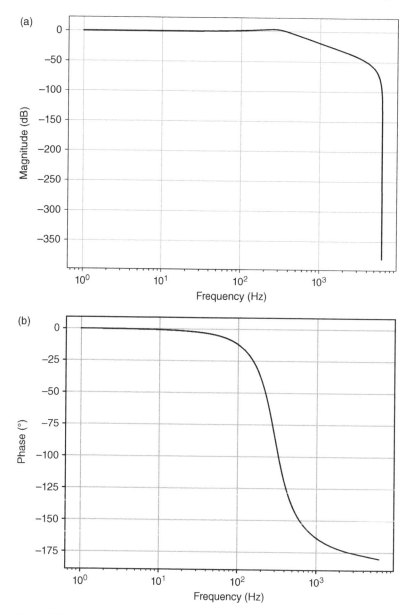

Figure 8.7 Bode plot of the DC motor after applying the PI controller to it. (a) Magnitude plot. (b) Phase plot.

```
num4 = [Kd, 0]
den4 = [1]

cds = mpcontrolPC.tf(num4, den4)

#Digital PID Controller
Ts = 0.0005

cpz = mpcontrolPC.c2d(cps, Ts)
ciz = mpcontrolPC.c2d(cis, Ts)
cdz = mpcontrolPC.c2d(cds, Ts)

cz1 = mpcontrolPC.parallel(cpz, ciz)
cz = mpcontrolPC.parallel(cz1, cdz)
```

As the Python code in Listing 8.6 is executed, we obtain PID controller gain parameters as $K_p = 0.8753$, $K_i = 152.0259$, and $K_d = 0.0013$. Based on these values and bilinear transformation, the PID controller in digital form becomes

$$C(z) = \frac{5.9528z^2 - 10.0030z + 4.2022}{z^2 - 1} \tag{8.6}$$

We can obtain unit-step response of the closed-loop system using the Python code in Listing 8.2. To do so, we should modify transfer function of the controller as num2 = [5.9528, -10.0030, 4.2022] and den2 = [1, 0, -1]. We plot the obtained response in Figure 8.8. We can also calculate the closed-loop system time domain performance criteria using the Python code in Listing 8.3. To do so, we should modify transfer function of the controller as num2 = [5.9528,

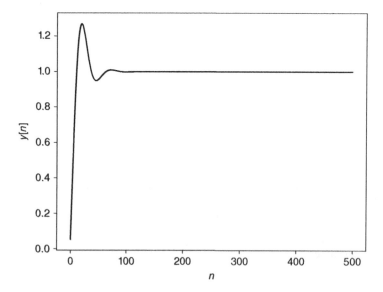

Figure 8.8 Unit step response of the closed-loop system formed by the PID controller.

-10.0030, 4.2022] and den2 = [1, 0, -1]. We will analyze the time domain performance criteria in a comparative way in Section 8.2.3.

We can analyze the closed-loop system formed by the PID controller in frequency domain. To do so, we should modify the transfer function of the controller as num2 = [5.9528, -10.0030, 4.2022] and den2 = [1, 0, -1] in the Python code in Listing 8.4. As we execute this code, we obtain Bode plot of the system both in magnitude and phase terms for 1000 elements as in Figures 8.9(a) and 8.9(b), respectively. As in the P and PI controllers, we can obtain the frequency domain performance criteria by modifying the Python code in Listing 8.4. This way, we will analyze the effect of the PID controller in frequency domain in a comparative way in Section 8.2.3.

8.2.3 Comparison of the Designed P, PI, and PID Controllers

In this section, we will first analyze the effect of designed controllers on time domain performance criteria. Therefore, we tabulate the T_{rt}, M_p, T_{pt}, T_{st}, and e_{ss} values obtained when the P, PI, and PID controllers are applied to the DC motor in Table 8.6. As can be seen in this table, T_{rt}, T_{pt}, T_{st}, and e_{ss} values decreased when we use P, PI, and PID controller compared to no controller scenario. On the other hand, these controllers increased the M_p value.

We can analyze Table 8.6 in more detail. As can be seen in this table, the P controller aims to reduce e_{ss} in the closed-loop system quickly. However, the possibility of oscillation also increases at output. Therefore, K_p should not be set to a high value. We can also observe from Table 8.6 that adding the integral controller tends to increase both M_p and T_{st}. On the other hand, it reduces e_{ss}. We can notice that the proportional gain, K_p, in the PI controller is reduced as desired. The integral controller in the PI controller eliminates e_{ss}.

We will next analyze the effect of designed controllers on frequency domain performance criteria. Therefore, we tabulate the GM, PM, ω_{cg}, and ω_{cp} values obtained when the P, PI, and PID controllers are applied to the DC motor in Table 8.7. As a reminder, when PM decreases, ζ decreases. The decrease in ζ increases M_p as mentioned in Chapter 7. While the integrator action in PI controller reduces e_{ss}, it increases both M_p and T_{st}. Hence, PM is reduced in the PI controller case (compared to the P controller) and the closed-loop system becomes more oscillatory. The derivative term in the PID controller tends to increase PM. Hence, the oscillation is reduced (less M_p compared to the PI controller). Furthermore, this improves GM. As a reminder, higher GM leads to a more stable system. The system with PID controller has the highest GM. Therefore, the closed-loop system formed by the PID controller is more stable compared to the ones constructed by the P and PI controllers.

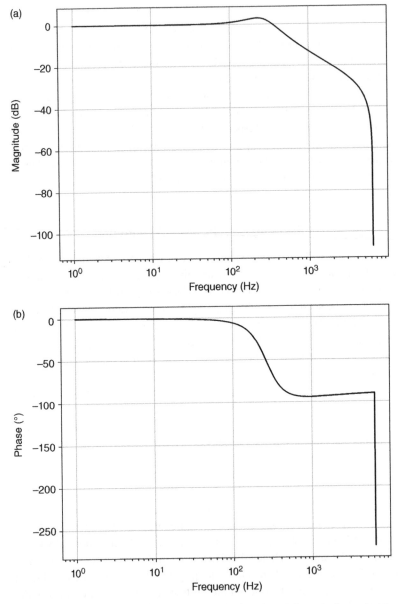

Figure 8.9 Bode plot of the DC motor after applying the PID controller to it. (a) Magnitude plot. (b) Phase plot.

Table 8.6 Comparison of the P, PI, and PID controllers by the time domain performance criteria.

System	T_{rt} (sec)	M_p (%)	T_{pt} (sec)	T_{st} (sec)	e_{ss}
No controller	0.0485	0	0.4345	0.0880	−9.4778
With P controller	0.0045	14.2537	0.0100	0.0205	0.1157
With PI controller	0.0045	35.4436	0.0105	0.0280	−3.8968e−13
With PID controller	0.0040	27.0234	0.0095	0.0285	−4.0902e−14

Table 8.7 Comparison of the P, PI, and PID controllers by the frequency domain performance criteria.

System	GM (dB)	PM (°)	ω_{cg} (rad/sec)	ω_{cp} (rad/sec)
No controller	313.071	53.094	6283.190	338.802
With P controller	313.071	∞	6283.190	Unavailable
With PI controller	378.426	64.781	6283.190	373.848
With PID controller	∞	105.343	Unavailable	352.644

8.3 Lag–Lead Controller Structure

The lag–lead structure provides another option to design a controller. Transfer function of a lag or lead controller is

$$C(z) = K_d \frac{1 - z_o}{1 - z_p} \tag{8.7}$$

where z_p and z_o represent pole and zero of the controller, respectively. K_d is the constant gain of the controller calculated according to $\frac{1-z_p}{1-z_o}$. The zero is less than the pole ($z_o < z_p$) and the gain K_d is less than one for the lag controller. The controller is of lead type if the zero is greater than the pole ($z_o > z_p$) and the gain K_d is greater than one. We will introduce design methods for both controllers and their combination as the lag–lead controller in the following sections.

8.3.1 Lag Controller

The lag controller improves transient-response of the overall system and its e_{ss} by adding a zero and pole to it. The closed-loop control scheme for the lag controller is as in Figure 8.10. It is generally advisable to select pole, z_p, and zero, z_o, of the lag controller close together.

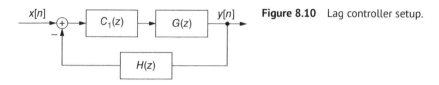

Figure 8.10 Lag controller setup.

8.3.2 Lead Controller

The lead controller has similar structure as the lag controller. It permits us to design a controller that improves the overall system stability and increases speed of the response. The closed-loop control scheme for the lead controller is as in Figure 8.11.

The system will approach the desired value in less time when the lead controller is added to it. Moreover, the lead controller adds positive phase to the system. When a positive phase is added to the system, its PM increases. When the PM increases, the ζ of the system increases. The increase in ζ reduces M_p. Furthermore, ω_{cg} increases which results in reduced T_{rt} and T_{st}.

8.3.3 Lag–Lead Controller

The lag–lead controller is formed by connecting the lag and lead controllers in series form. Hence, this controller merges the effect of lag and lead controllers such that we will have an overall system with improved transient response and stability. The closed-loop control scheme for the lag–lead controller is as in Figure 8.12.

To design a lag–lead controller in time domain, it is generally suggested to design the lead controller to achieve the desired transient response and stability requirements. Then, the lag controller should be designed to improve the steady-state response of the lead controlled system. To do so, we should first evaluate the performance of the closed-loop system without any controller by observing T_{rt}, M_p,

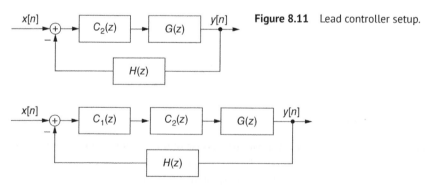

Figure 8.11 Lead controller setup.

Figure 8.12 Lag–lead controller setup.

T_{pt}, and T_{st}. Then, we should design the lead controller to meet the desired transient response specifications by selecting the z_p and z_o values given in Section 8.3.2. Afterward, we should analyze closed-loop response of the system with the lead controller and evaluate its e_{ss} performance. Finally, we should determine how much improvement in e_{ss} is required and select the z_p and z_o values of lag controller given in Section 8.3.1 accordingly.

To design a lag–lead controller in frequency domain, it is generally suggested to design the lead controller first. During this phase, time domain performance and stability-based design criteria should be considered. The lead controller in this setup increases PM (hence reduces M_p). Then, the lag controller should be designed. Here, the aim is to improve e_{ss} of the overall system.

8.4 Lag–Lead Controller Design in MATLAB

We will benefit from MATLAB to design the lag, lead, and lag–lead controllers in this book. To do so, we will use the Control System Designer (CSD) tool which will allow us to design a controller satisfying design constraints.

8.4.1 Control System Designer Tool

We can use the CSD tool under MATLAB to design a controller considering specific requirements in time or frequency domain using a graphical user interface. This tool allows modifying the control structure and observing the system response interactively. It is possible to add, modify, and remove controller poles, zeros, and gain with the tool. Furthermore, it allows us to observe the time and frequency domain response of the system.

The assumed controller structure in the CSD tool is as in Figure 8.13. Here, only the controllers $G(z)$ and $C(z)$ can be designed. The transfer function $H(z)$ in the feedback loop has value one. In other words, the CSD tool assumes a unity feedback in operation.

The CSD tool can be started from the MATLAB command window by executing the command `controlSystemDesigner(g)`. Here, g is transfer function of the system to be controlled. Before starting the tool, we should form it. Let us take the digital representation of our DC motor with the transfer function given in Eq. (6.18). We can define g and start the CSD tool under MATLAB with the following code snippet.

```
Ts=0.0005;

num = [0.00943, 0.01886, 0.00943];
den  = [1, -1.8188, 0.8224];

g=tf(num,den, Ts);
controlSystemDesigner(g);
```

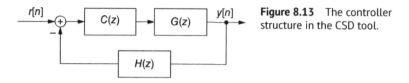

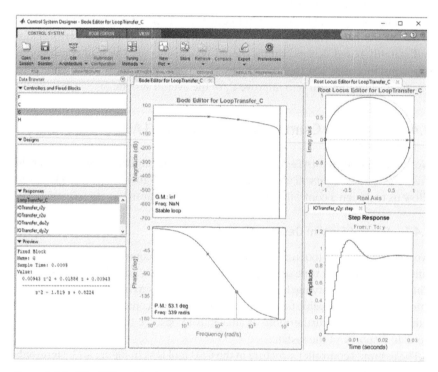

Figure 8.13 The controller structure in the CSD tool.

Figure 8.14 The CSD tool interactive window. (Source: Matlab, The MathWorks, Inc.)

As the CSD tool starts, it will open an interactive window as in Figure 8.14. This window initially summarizes all necessary information about the system to be controlled. To be more specific, the reader can analyze the system characteristics in frequency domain (as Bode and Nyquist plots), complex plane (as root-locus plot), and in time-domain (as step response plot) here.

As mentioned before, the CSD tool window is interactive. Hence, the reader can observe the system characteristics from the selected plot. We can show this by focusing on the step response plot of the DC motor. As we right click on the step response plot, a new pop up window appears as in Figure 8.15. The reader can select the peak response (M_p), settling time (T_{st}), rise time (T_{rt}), and steady-state value of the system from there. Afterward, the reader can move

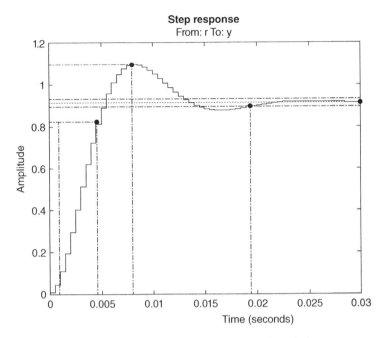

Figure 8.15 Step response system parameters selection window.

the cursor above the selected characteristic and observe the value there. For our DC motor, these values are obtained from the tool as $T_{rt} = 0.0037$ seconds, $M_p = 20.0\%$, $T_{pt} = 0.008$ seconds, $T_{st} = 0.0194$ seconds, and the steady-state value being equal to 0.087. To note here, MATLAB initially assumes $C(z) = 1$ and $H(z) = 1$ in Figure 8.13. Hence, the obtained results are for the constructed closed-loop system accordingly.

The reader can also use the CSD tool to design a controller for the system at hand by setting the required criteria in time or frequency domain. We will focus on the controller design via root-locus and Bode plots in this book. Let us start with the root-locus plot based controller design next.

8.4.2 Controller Design in Complex Plane

We can use the root-locus plot to generate a digital controller for our system. We can use the CSD tool for this purpose. To do so, we should execute the command `controlSystemDesigner('rlocus', g)` in the MATLAB command window. Different from the previous section, here we specify the design method as `rlocus`. As we execute the command, we will have the CSD tool window as in Figure 8.16.

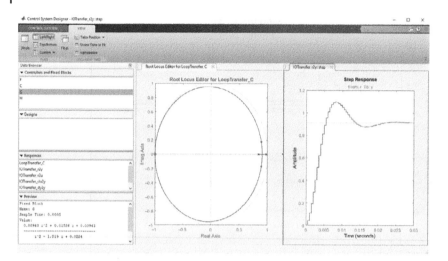

Figure 8.16   Root-locus editor in the CSD tool. (Source: Matlab, The MathWorks, Inc.)

We can design the controller by adding design requirements to the interactive root-locus editor. To do so, we should right click on the editor and select the "Design Requirements" option from the pop-up list. As we select "New" there, a new pop-up window, titled "New Design Requirements," appears asking for the specific requirements as "settling time," "percent overshoot," "damping ratio," "natural frequency," and "region constraint." Assume that we would like to design a lead controller for our DC motor to set $T_{st} < 0.02$ seconds and $M_p < 10.0\%$. To do so, we should enter these values step by step. In other words, we should first enter the settling time to be less than 0.02 seconds from the "New Design Requirements" pop-up window. As we enter this value, the interactive root-locus editor will have two regions as yellow and white. This indicates that the controller poles and zeros should be in the white region to satisfy this design constraint. We can click on the white region and add the next design constraint as $M_p < 10.0\%$ from the "New Design Requirements" pop-up window. Then, possible pole locations will be shown as new white region. We provide this region for our example in Figure 8.17.

As the possible region to satisfy the design constraints is determined, the user can right click on it and select the "Add Pole Zero" option from the pop-up window. Here, we can select adding poles, zeros, or lead, lag, or notch type controller.

8.4.2.1   Lag Controller

For our case, we will select the lag controller first. Afterward, as we press on a point in the allowable region of the root-locus plot, the CSD tool designs the corresponding controller and provides it in the leftmost "Preview" window. The reader should select the controller "C" from the "Controllers and Fixed Blocks" subwindow. The

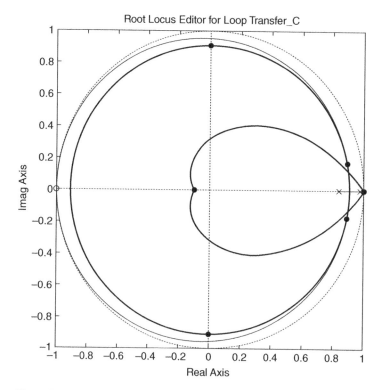

Figure 8.17 Possible pole locations in the root-locus editor for our controller.

designed lag controller after all these operations will be as

$$C_1(z) = 0.6514 \frac{z - 0.9950}{z - 1} \tag{8.8}$$

To note here, the reader can obtain a different controller from the CSD tool when a different point is selected in the interactive plot. Therefore, the lag controller in Eq. (8.8) is not unique.

We can implement the designed lag controller in Eq. (8.8) in Python. Afterward, we can obtain unit-step response of the closed-loop system using the Python code in Listing 8.2. To do so, we should modify the transfer function of the controller as num2 = [0.6514, -0.6514*0.9950] and den2 = [1, -1]. We plot the obtained response in Figure 8.18. We can have some key observations from this figure. However, the better way is tabulating the system performance criteria in time domain. To do so, we should modify the Python code in Listing 8.3 by replacing num2 = [0.6514, -0.6514*0.9950] and den2 = [1, -1]. We will analyze the time domain performance criteria in a comparative way in Section 8.4.2.4.

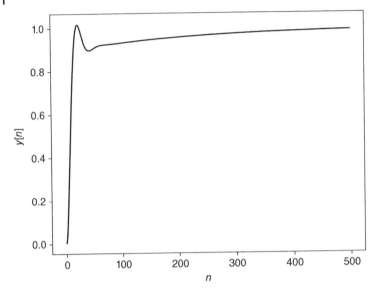

Figure 8.18 Unit step response of the closed-loop system formed by the lag controller designed in MATLAB.

In order to analyze the closed-loop system formed by the lag controller in frequency domain, we should first modify the transfer function of the controller as `num2 = [0.6514, - 0.6514*0.9950]` and `den2 = [1, -1]` in Listing 8.4. As we obtain the frequency domain performance criteria, we will analyze the effect of the lag controller in frequency domain in a comparative way in Section 8.4.2.4.

8.4.2.2 Lead Controller

We can repeat the same steps as in the lag controller to design the lead controller. To do so, we should select the lead controller at the "Add Pole or Zero" step. As we proceed, we will obtain the lead controller designed by MATLAB as

$$C_2(z) = 2.6613 \frac{z - 0.9}{z - 0.7339} \tag{8.9}$$

To note here, the reader can obtain a different controller from the CSD tool when a different point is selected in the interactive plot. Therefore, the lead controller in Eq. (8.9) is not unique.

We can implement the designed lead controller in Eq. (8.9) in Python. Afterward, we can obtain unit-step response of the closed-loop system using the Python code in Listing 8.2. To do so, we should modify transfer function of the controller as `num2 = [2.6613, -2.6613*0.9]` and `den2 = [1, -0.7339]`. We plot the unit step response in Figure 8.19. By updating the Python code in Listing 8.3 as

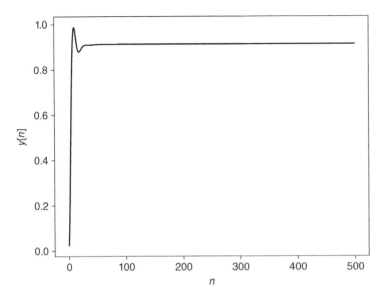

Figure 8.19 Unit step response of the closed-loop system formed by the lead controller designed in MATLAB.

`num2 = [2.6613, -2.6613*0.9]` and `den2 = [1, -0.7339]`, we can obtain the time domain performance criteria for the closed-loop system formed by the lead controller. We will analyze these in a comparative way in Section 8.4.2.4.

We can also analyze the closed-loop system formed by the lead controller in frequency domain. To do so, we should modify the transfer function of the controller in Listing 8.4 as `num2 = [2.6613, -2.6613*0.9]` and `den2 = [1, -0.7339]`. As we obtain the frequency domain performance criteria, we will analyze the effect of the lag controller in frequency domain in a comparative way in Section 8.4.2.4.

8.4.2.3 Lag–Lead Controller

We can merge the lag and lead controllers designed in MATLAB to form the final lag–lead controller. To do so, we should connect the controllers given in Eqs. (8.8) and (8.9) in series form. We can realize this by the function `series` in the MicroPython control systems library. The final lag–lead controller will be as

$$C(z) = \frac{1.7336z^2 - 3.2851z + 1.5524}{z^2 - 1.7339z + 0.7339} \tag{8.10}$$

We can implement the designed lag–lead controller in Eq. (8.10) in Python. Afterward, we can obtain unit-step response of the closed-loop system using the Python code in Listing 8.7. We plot the response in Figure 8.20.

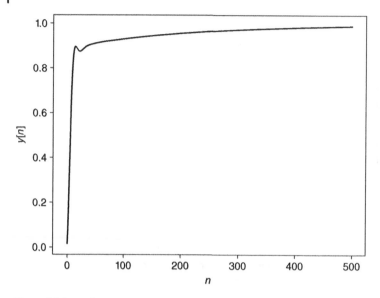

Figure 8.20 Unit step response of the closed-loop system formed by the lag–lead controller designed in MATLAB.

Listing 8.7: Unit step response of the closed-loop system formed by the lag–lead controller for the DC motor.

```
import mpcontrol
import mpcontrolPC

#System
Ts = 0.0005
num1 = [0.00943, 0.01886, 0.00943]
den1 = [1, -1.8188, 0.8224]

#Transfer function of the system
g = mpcontrolPC.tf(num1,den1,Ts)

#lag controller
num2 = [0.6514, -0.6514*0.995]
den2 = [1, -1]
c1 = mpcontrolPC.tf(num2,den2,Ts)

#lead controller
num3 = [2.6613, -2.6613*0.9]
den3 = [1, -0.7339]
c2 = mpcontrolPC.tf(num3,den3,Ts)

#lag-lead controller
c = mpcontrolPC.series(c1,c2)

#Transfer function of the feedback system
num4 = [1]
den4 = [1]
```

```
h = mpcontrolPC.tf(num4,den4,Ts)

# Transfer function of the closed-loop controlled system
gol = mpcontrolPC.series(g,c)
gcl = mpcontrolPC.feedback(gol,h)

#Step input signal
N = 500
x = mpcontrol.step_signal(N, 1)

#Calculate the output signal
y = mpcontrol.lsim(gcl, x)
```

In order to have a quantitative comparison of the designed lag–lead controller with the lag and lead controllers, we should add the `gcl.stepinfo()` and `gcl.ss_error('Step')` functions to the Python code in Listing 8.7. We will analyze the time domain performance criteria in a comparative way in Section 8.4.2.4.

We can also analyze the closed-loop system formed by the lag–lead controller in frequency domain. To do so, we can use the Python code in Listing 8.8 on PC. As we obtain the frequency domain performance criteria, we will analyze the effect of the lag controller in frequency domain in a comparative way in Section 8.4.2.4.

Listing 8.8: Frequency domain parameters of the closed-loop system formed by the lag–lead controller for the DC motor.

```
import mpcontrolPC

#System
Ts = 0.0005
num1 = [0.00943, 0.01886, 0.00943]
den1 = [1, -1.8188, 0.8224]

#Transfer function of the system
g = mpcontrolPC.tf(num1,den1,Ts)

#lag controller
num2 = [0.6514, -0.6514*0.995]
den2 = [1, -1]
c1 = mpcontrolPC.tf(num2,den2,Ts)

#lead controller
num3 = [2.6613, -2.6613*0.9]
den3 = [1, -0.7339]
c2 = mpcontrolPC.tf(num3,den3,Ts)

#lag-lead controller
c = mpcontrolPC.series(c1,c2)

#Transfer function of the feedback system
num4 = [1]
den4 = [1]

h = mpcontrolPC.tf(num4,den4,Ts)
```

```
# Transfer function of the closed-loop controlled system
gol = mpcontrolPC.series(g,c)
gcl = mpcontrolPC.feedback(gol,h)

#Bode plot of the System
[mag, phase, freq] = mpcontrolPC.bode(gcl,1000)

#GM, PM, Crossover frequency values of the system
mpcontrolPC.margin(gcl)
```

8.4.2.4 Comparison of the Designed Lag, Lead, and Lag–Lead Controllers

In this section, we will first analyze the effect of designed controllers on time domain performance criteria. Therefore, we tabulate the T_{rt}, M_p, T_{pt}, T_{st}, and e_{ss} values for the lag, lead, and lag–lead controller based closed-loop systems for the DC motor in Table 8.8. As can be seen in this table, both design requirements ($T_{st} < 0.02$ and $M_p < 10.0\%$) are satisfied only by the lead controller. $M_p < 10.0\%$ requirement is satisfied by the lag and lag–lead controllers. We furthermore notice that the lag controller improves e_{ss}. However, the T_{st} value also increases with this controller. Thus, design requirement $T_{st} < 0.02$ could not be satisfied. We can design a lag controller that satisfies the $T_{st} < 0.02$ requirement. However, in this case, we obtain a system response with higher M_p and e_{ss} values. As a result, we can deduce that the lag controller improves e_{ss}, but reduces speed of the response. The lead controller tends to increase speed of the response. This means that the system will approach to the desired value in less time (as can be seen in less T_{rt} and T_{st}). However, e_{ss} value in the lead controller becomes larger than the lag controller case. If both smaller M_p and small e_{ss} is required, lag–lead controller is the best option since it aims to improve both transient response and steady-state error at the same time.

Table 8.9 summarizes frequency domain comparison results for the lag, lead, and lag–lead controllers. As can be seen in this table, PM of the system with all controllers is infinite which indicates the system is robust and has minimal M_p. As a reminder, a high GM results in more stable system. Thus, the stability is reduced when we use the lag–lead controller.

Table 8.8 Comparison of the lag, lead, and lag–lead controllers by the time domain performance criteria.

System	T_{rt} (sec)	M_p (%)	T_{pt} (sec)	T_{st} (sec)	e_{ss}
No controller	0.0485	0	0.4345	0.0880	−9.4778
With lag controller	0.0060	1.716	0.0105	0.1855	2.7278e−12
With lead controller	0.0020	8.056	0.0050	0.0115	0.0871
With lag–lead controller	0.0175	0	2.2305	0.1890	1.3805e−11

Table 8.9 Comparison of the lag, lead, and lag–lead controllers by the frequency domain performance criteria.

System	GM (dB)	PM (°)	ω_{cg} (rad/sec)	ω_{cp} (rad/sec)
No controller	313.071	53.094	6283.190	338.802
With lag controller	684.000	∞	6283.190	Unavailable
With lead controller	672.035	∞	6283.190	Unavailable
With lag–lead controller	371.080	∞	6283.190	Unavailable

8.4.3 Controller Design in Frequency Domain

The lag, lead, and lag–lead controllers considered in the previous section can also be designed in frequency domain. To do so, we benefit from the Bode plot. Besides, system requirements should be represented in frequency domain metrics as PM, GM, and bandwidth. We will next introduce design procedures for the lag, lead, and lag–lead controllers in frequency domain.

We can use the interactive CSD tool to design our controller in frequency domain by executing the command `controlSystemDesigner('bode', g)`. As can be seen here, we specified the design method as `bode`. As we execute the command, we will have the CSD tool window as in Figure 8.21.

The reader can observe the GM, PM, ω_{cg}, and ω_{cp} values of the system to be controlled in the interactive Bode-plot editor. These values are as GM = ∞ dB, PM = 53.1°, ω_{cg} undefined, and ω_{cp} = 339 rad/s for our DC motor. To note here, MATLAB initially assumes $C(z) = 1$ and $H(z) = 1$ in Figure 8.13. Hence, the obtained results are for the corresponding closed-loop system.

We can design the controller by adding design requirements to the interactive Bode plot editor. To do so, we should right click on the editor and select the "Design Requirements" option from the pop-up list. As we select "New" there, a new pop-up window, titled "New Design Requirements," appears asking for the specific requirements as "Upper gain limit," "Lower gain limit," and "Gain & Phase margins."

Assume that we would like to design a controller for our DC motor to set GM > 20 dB and PM > 70°. We can enter both values from the "New Design Requirements" list "Gain & Phase margins" option from the CSD window in Figure 8.21. We can design the controller using the interactive Bode plot by right clicking on it and selecting the "Add Pole Zero" option from the pop-up window. Here, we can select adding poles, zeros, or lead, lag, or notch type controller.

8.4.3.1 Lag Controller

As the first option, we will select the lag controller for design. Then, the CSD tool asks to place the lag controller on Bode plot. As we press the left button, it places

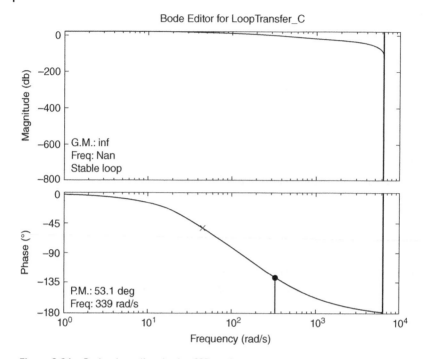

Figure 8.21 Bode plot editor in the CSD tool.

the controller pole and zero. At the same time, it provides the gain and phase margins obtained. If the initial location of the zero or pole is not suitable, the reader can move the zero or pole by left clicking on it. Hence, he or she can design the controller interactively. The reader can also observe the effect of moving the pole or zero in time domain, by the step response as well. Afterward, the CSD tool designs the corresponding controller and provides it in the leftmost "Preview" window. The reader should select the controller "C" from the "Controllers and Fixed Blocks" subwindow. The designed lag controller after all these operations will be as

$$C_1(z) = 0.4352 \frac{z - 0.9909}{z - 0.9960} \tag{8.11}$$

To note here, the reader can obtain a different controller from the CSD tool when a different point is selected in the interactive plot. Therefore, the lag controller in Eq. (8.11) is not unique.

We can implement the designed lag controller in Eq. (8.11) in Python. Afterward, we can analyze the closed-loop system formed by this controller in frequency domain. To do so, we should modify transfer function of the controller in Listing 8.4 as num2 = [0.4352, -0.4352*0.9909] and den2 = [1, -0.996]. As we execute this code, we obtain Bode plot of the system both in magnitude and phase terms for 1000 elements as in Figures 8.22(a) and 8.22(b), respectively.

Although Figure 8.22 is helpful in understanding the effect of the lag controller on the DC motor, it is better to obtain frequency domain performance criteria to quantify this effect. Therefore, we can add the function `margin()` to the Python code in Listing 8.4. As we obtain the frequency domain performance criteria, we will analyze the effect of the lag controller in frequency domain in a comparative way in Section 8.4.3.4.

8.4.3.2 Lead Controller

We can repeat the same steps as in the lag controller to design the lead controller in frequency domain. To do so, we should select the lead controller at the "Add Pole or Zero" step. As we proceed, we will obtain the lead controller designed by MATLAB as

$$C_2(z) = 3.1574 \frac{z - 0.8668}{z - 0.5793} \tag{8.12}$$

To note here, the reader can obtain a different controller from the CSD tool when a different point is selected in the interactive plot. Therefore, the lead controller in Eq. (8.12) is not unique.

In order to analyze the effect of the designed lead controller in the closed-loop system, we should modify the transfer function of the controller in Listing 8.4 as `num2 = [3.1574, -3.1574*0.8668]` and `den2 = [1, -0.5793]`. As we execute this code, we obtain Bode plot of the system both in magnitude and phase terms for 1000 elements as in Figures 8.23(a) and 8.23(b), respectively. We will also analyze the effect of lead controller in frequency domain in a comparative way in Section 8.4.3.4.

8.4.3.3 Lag–Lead Controller

We can merge the lag and lead controllers designed in MATLAB to form the final lag–lead controller. To do so, we should connect the controllers given in Eqs. (8.11) and (8.12) in series form. We can realize this by the function `series` in the MicroPython control systems library. The final lag–lead controller will be as

$$C(z) = \frac{1.3740z^2 - 2.5526z + 1.1802}{z^2 - 1.5753z + 0.5770} \tag{8.13}$$

We should modify transfer function of the lag and lead controllers in Listing 8.8 as `num2 = [0.4352 , -0.4352*0.9909]`, `den2 = [1, -0.996]` and `num3 = [3.1574, -3.1574*0.8668]`,`den3 = [1, -0.5793]` in order to implement the designed lag–lead controller. As we execute this code, we obtain Bode plot of the system both in magnitude and phase terms for 1000 elements as in Figures 8.24(a) and 8.24(b), respectively. We will also analyze the effect of the lag–lead controller in frequency domain in a comparative way in Section 8.4.3.4.

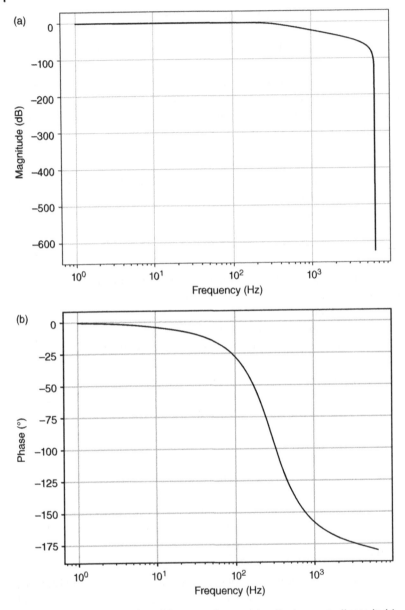

Figure 8.22 Bode plot of the DC motor after applying the lag controller to it. (a) Magnitude plot. (b) Phase plot.

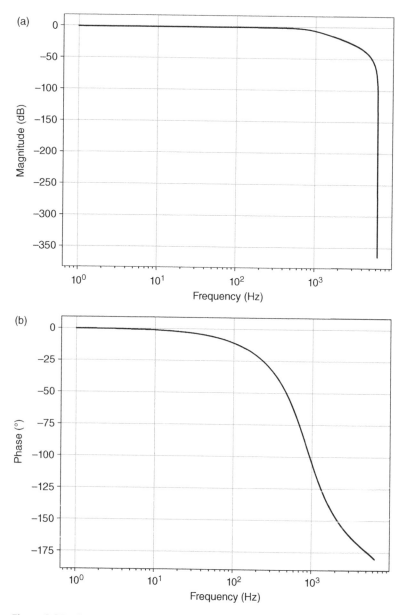

Figure 8.23 Bode plot of the DC motor after applying the lead controller to it. (a) Magnitude plot. (b) Phase plot.

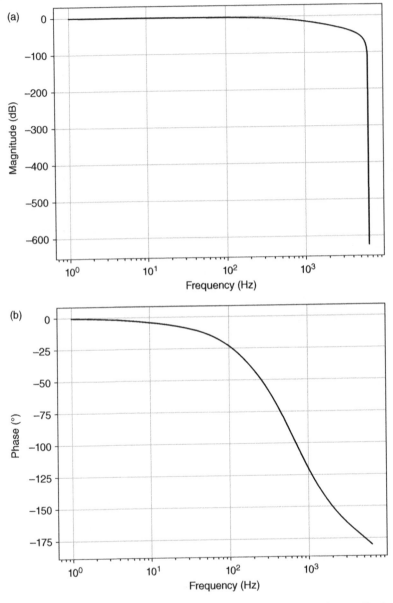

Figure 8.24 Bode plot of the DC motor after applying the lag–lead controller to it. (a) Magnitude plot. (b) Phase plot.

Table 8.10 The effect of lag, lead, and lag–lead controllers on system parameters in frequency domain.

System	GM (dB)	PM (°)	ω_{cg} (rad/sec)	ω_{cp} (rad/sec)
No controller	313.071	53.094	6283.190	338.802
With lag controller	685.479	∞	6283.190	Unavailable
With lead controller	364.134	∞	6283.190	Unavailable
With lag–lead controller	313.071	∞	6283.190	Unavailable

8.4.3.4 Comparison of the Designed Lag, Lead, and Lag–Lead Controllers

Table 8.10 summarizes frequency domain comparison results for the lag, lead, and lag–lead controllers. As can be seen in this table, we satisfy the design requirement of the GM > 20 dB and PM > 70° in all lag, lead, and lag–lead controllers. PM of the system with all controllers are infinite which indicates the system is stable.

8.5 Application: Adding Closed-Loop Digital Controller to the DC Motor

This is the extended version of the application introduced in Chapter 7. Here, we improve the control action by adding a closed-loop controller to the system. Besides, the system setup for this application is the same as in Figure 7.11.

8.5.1 Hardware Setup

The hardware setup used in this application is the same as in Section 7.4. The only difference here is that we benefit from the encoder to form a feedback loop in the closed-loop control strategy. Besides, we also benefit from the encoder to send data to PC as well.

8.5.2 Procedure

Here, we form the application to control DC motor using closed-loop structure. To do so, we benefit from the PI and PID controllers obtained in Listing 8.5 and Listing 8.6, respectively. First, we apply a constant amplitude input with 60 rpm to the DC motor using these controllers. Then, we obtain speed output from the quadrature encoder and send data to PC in real-time for one second. Next, we use digital transfer function of the DC motor obtained in Chapter 6. We simulate the system with closed-loop controller by applying the same input signal. Finally, we compare the actual and simulated outputs in terms of unit step response plots and time domain performance criteria.

8.5.3 C Code for the System

We form a new Mbed Studio project to integrate the closed-loop controller to the DC motor. The software setup will be the same as in Section 3.4.3. Within the project, the main file will be as

```
int main()
{
    float num[2] = {0.675471, -0.637465};
    float den[2] = {1, -1};
//  float num[3] = {5.95279, -10.003, 4.20221};
//  float den[3] = {1, 0, -1};
    user_button_init();
    system_input_init(1, 12/3.3);
    system_output_init(1, 3591.84);
    send_data_RT(2);
    sampling_timer_init();
    controller_tf_init(num, den, 2, 2, 1);
//  controller_tf_init(num, den, 3, 3, 1);
    menu_init();
    while (true) {
        system_loop();
    }
}
```

Within `main`, the `select` input of the function `controller_init` is set as 1 to implement the closed-loop controller. The reader can replace the commented lines with their counterparts to implement the PID controller instead of the PI controller.

In order to compare the actual system output and simulation results, the reader should run the Python code in Listing 8.9 on PC. Then, the user button on the STM32 board should be pressed. Afterward, the DC motor starts running and data transfer to PC begins from the STM32 microcontroller.

Listing 8.9: The Python code to compare the actual system output and simulation results.

```
import mpcontrol
import mpcontrolPC
import matplotlib.pyplot as plt

N = 2000
signal1, signal2 = mpcontrolPC.get_data_from_MC(N, 'com10', 1)
mpcontrolPC.save_to_file(signal1, file_name='input.dat')
mpcontrolPC.save_to_file(signal2, file_name='output.dat')
```

```
signal3 = mpcontrolPC.ma_filter(signal2, 15)
mpcontrolPC.save_to_file(signal3, file_name='output_filtered.dat')

Ts = 0.0005
num = [165702.47]
den = [1, 390.2, 15701]

sys = mpcontrolPC.tf(num, den)
sys_d = mpcontrolPC.c2d(sys, Ts)

[Kp, Ki, Kd] = mpcontrolPC.pid_tune(sys, 'ZieglerNichols', 'PI')
num2 = [Kp]
den2 = [1]
cps = mpcontrolPC.tf(num2, den2)

num3 = [Ki]
den3 = [1, 0]
cis = mpcontrolPC.tf(num3, den3)

#num4 = [Kd, 0]
#den4 = [1]
#cds = mpcontrolPC.tf(num4, den4)

cpz = mpcontrolPC.c2d(cps, Ts)
ciz = mpcontrolPC.c2d(cis, Ts)
#cdz = mpcontrolPC.c2d(cds, Ts)
#cz1 = mpcontrolPC.parallel(cpz, ciz)
#cz = mpcontrolPC.parallel(cz1, cdz)
cz = mpcontrolPC.parallel(cpz, ciz)

tf_forward = mpcontrolPC.series(cz, sys_d)

num5 = [1]
den5 = [1]

h = mpcontrolPC.tf(num5,den5,Ts)
gcl = mpcontrolPC.feedback(tf_forward,h)

y = mpcontrolPC.compare_stepinfo(signal3, gcl, 60)
mpcontrolPC.save_to_file(y, file_name='output_simulation.dat')
```

In Listing 8.9, we should set `get_data_from_MC(NoElements, com_port, selection)` for real-time data transfer for two signals. Then, the function `save_to_file()` is used to save received data to a file on PC. We construct digital transfer function of the DC motor with closed-loop controller. We then obtain time domain performance criteria for both simulation and actual output data using `compare_stepinfo(Real_Time_Output, Transfer_Function, Step_Gain)`.

8.5.4 Python Code for the System

We can repeat the application by embedding the Python code on the STM32 microcontroller. To do so, we should replace the content of the `main.py` file with the code in Listing 8.10 by following the steps explained in Section 3.2.2. We should also embed the `system_general.py`, `system_input.py`, system_output.py, system_controller.py, and mpcontrol.py files to flash. To note here, these files are provided in the accompanying book web site. As a reminder, the Python code to be used on the PC side is the same as in Section 8.5.3.

Listing 8.10: The Python code to be used in closed-loop controller for the DC motor.

```
import system_general
import system_input
import system_output
import system_controller

def main():
        system_general.button_init()
        system_general.PWM_frequency_select(0)
        system_general.sampling_timer_init(0)
        system_general.input_select(input_type='Constant Amplitude',
           amplitude=60)
        system_controller.controller_init([0.675471, -0.637465],
           [1, -1], 0.0005, 1)
#       system_controller.controller_init([5.95279, -10.003,
    4.20221], [1, 0, -1], 0.0005, 1)
        system_general.motor_direction_select(1)
        system_input.system_input_init(2, 3.63636363636)
        system_output.system_output_init(1, 3591.84)
        system_general.send_data_RT(2)
        while True:
                system_general.system_loop()
main()
```

8.5.5 Observing Outputs

When we execute the C code of the closed-loop PID controller, the resulting signals will be as in Figure 8.25. We can obtain similar results when we execute the Python code for this application as well. As can be seen in this figure, output oscillates around 60 rpm and does not settle down. The main reason for this is that data obtained from the encoder is noisy and it is amplified by the derivative term in the PID controller. Therefore, the actual output cannot settle down. Hence, we will use the PI controller instead of PID.

We can repeat the same process when the PI controller is used in the closed-loop strategy. We provide the actual and simulated output signals obtained from the C code this way in Figure 8.26. We can obtain similar results when we execute the Python code for this application as well. As can be seen in this figure, e_{ss} is decreased for both actual and simulation outputs compared to the open-loop control strategy in Section 7.4.

As can be seen in Figure 8.25, transient response for the actual and simulated systems are different. The main reason for this difference is that we can only apply a voltage value between 0 and 12 V to the DC motor in practice. The PI parameters

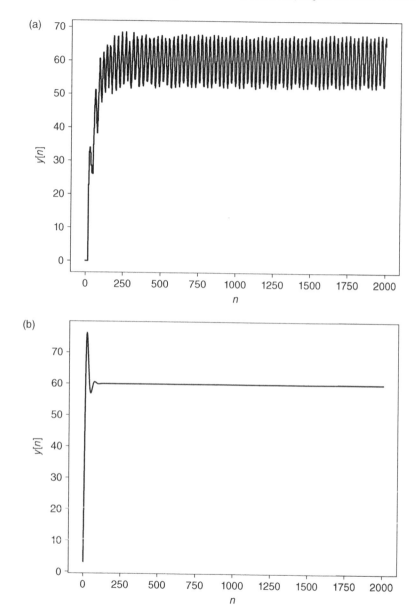

Figure 8.25 Actual (a) and simulated (b) output signals obtained from the DC motor with the PID controller.

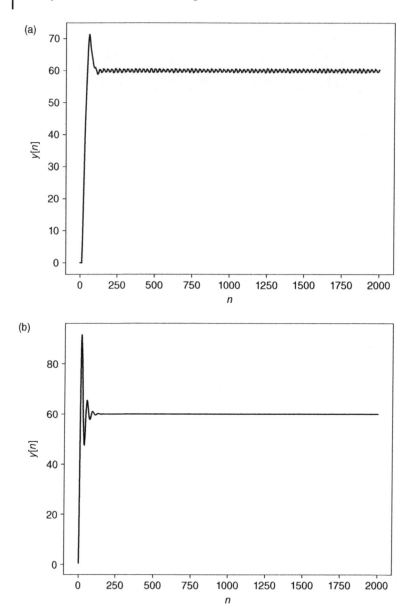

Figure 8.26 Actual (a) and simulated (b) output signals obtained from the DC motor with the PI controller.

Table 8.11 The time domain performance criteria comparison for the DC motor with the PI controller.

Measure	Actual	Simulation
Rise time (T_{rt}), sec	0.0115	0.0045
Settling time (T_{st}), sec	0.0450	0.0280
Peak time (T_{pt}), sec	0.0285	0.0105
Overshoot (M_p), %	18.9850	35.5308
Undershoot, %	0	0
Settling min.	55.3984	56.3300
Settling max.	71.3431	81.3185
Peak	71.3431	81.3185

are not calculated taking such a constraint into account. Moreover, the PI controller gives output above 12 V to decrease T_{rt}. Unfortunately, the actual system cannot generate such a signal.

We tabulate the time domain performance criteria from the actual and simulation outputs in Table 8.11. While calculating these values, the actual output data was filtered using a moving average filter to eliminate noise from it. The difference between the actual and simulation outputs mentioned in the previous paragraph can also be observed in Table 8.11.

8.6 Summary

This chapter focused on digital controller design methods based on transfer functions. Therefore, we first considered the PID controller and its design methodology. We provided functions in the MicroPython control systems library to design and construct the P, PI, or PID controller for a given system. We applied these controllers to the DC motor and compared the resultant systems in terms of their time and frequency domain performance criteria. We followed the same strategy to the lag, lead, and lag–lead controller design. However, we used the CSD tool under MATLAB for this purpose. As the end of chapter application, we applied the designed PI and PID controllers to the DC motor via the STM32 microcontroller and observed their working principles. The reader can benefit from the methods introduced in this chapter to design and implement transfer function based controllers to the system at hand.

Problems

8.1 Consider the DC motor with transfer function
$$G(s) = \frac{165702.47}{s^2 + 390.2s + 15701}$$

 a. design P, PI, and PID controllers for the DC motor using the Cohen–Coon method. Form a separate closed-loop system for each controller separately.
 b. Plot the unit step response for the closed-loop systems.
 c. Compare the closed-loop systems in terms of time domain performance criteria.
 d. Obtain Bode plots for the closed-loop systems.
 e. Compare the closed-loop systems in terms of frequency domain performance criteria.

8.2 Repeat Problem 8.1 now using the Chien–Hrones–Reswick method in the controller design.

8.3 Consider the DC motor in Problem 8.1,
 a. Convert it to digital form by the bilinear transformation method.
 b. Design lag, lead, and lag–lead controllers (in closed-loop form) for the digital system in time domain that improves M_p by at least 50% and T_{st} by at least 10%.
 c. Plot unit step response for the closed-loop systems formed by the designed controllers.
 d. Compare the closed-loop systems formed by the designed controllers in terms of time domain performance criteria.
 e. Obtain Bode plots for the closed-loop systems formed by the designed controllers.
 f. Compare closed-loop systems formed by the designed controllers in terms of frequency domain performance criteria.

8.4 Consider the DC motor in Problem 8.1,
 a. Convert it to digital form by the bilinear transformation method.
 b. Design lag, lead, and lag–lead controllers (in closed-loop form) for the digital system in frequency domain such that the overall system has GM > 40 dB.

c. Plot the unit step response for the closed-loop systems formed by the designed controllers.

d. Compare the closed-loop systems formed by the designed controllers in terms of time domain performance criteria.

e. Obtain Bode plots for the closed-loop systems formed by the designed controllers.

f. Compare the closed-loop systems formed by the designed controllers in terms of frequency domain performance criteria.

9

State-space Based Control System Analysis

State-space provides another option to analyze and design digital controllers besides transfer function based methods. Therefore, we will explore the state-space approach in detail here. To do so, we will start with the definition of state and emphasize its difference from the transfer function approach. Then, we will show how state-space representation of an LTI system can be constructed from its differential or difference equation in continuous- and discrete-time, respectively. We know that differential equations lead to transfer function representation in continuous-time from Chapter 6. Likewise, difference equations lead to transfer function representation in discrete-time as mentioned in Chapter 4. Hence, we will cover methods on converting state-space and transfer function representations. Finally, we will look at properties of a system from its state-space representation. While performing all these, we will benefit from MicroPython and Python control systems libraries. As the end of chapter application, we will simulate state-space representation of the DC motor.

9.1 State-space Approach

In order to grasp the concepts to be covered in this chapter, we should first explain what we understand from the state. Then, we should emphasize the difference of state-space approach from transfer function based methods. We will perform these in this section.

9.1.1 Definition of the State

States of the system can be represented as the smallest number of variables such that their knowledge at time t, together with the knowledge of the input at that

Embedded Digital Control with Microcontrollers: Implementation with C and Python,
First Edition. Cem Ünsalan, Duygun E. Barkana, and H. Deniz Gürhan.
© 2021 The Institute of Electrical and Electronics Engineers, Inc. Published 2021 by John Wiley & Sons, Inc.
Companion website: www.wiley.com/go/Unsalan/Embedded_Digital_Control_with_Microcontrollers

time, completely determine the system behavior for any time. In other words, states are the minimum set of parameters of a system which completely describe its status in time. This is related to the state variable method used in describing differential equations. In this approach, differential equation of a system is organized as a set of first-order differential equations in vector form.

Minimum number of states is equal to the minimum number of first-order differential equations required to describe system dynamics completely. In practice, the number of states is typically equals to the number of independent energy storage elements in the system. However, the choice of states is not unique. Besides, a state may not correspond to a physical entity. But, it may still satisfy the definition of being a state.

9.1.2 Why State-space Representation?

As mentioned in Chapter 6, transfer function of a continuous-time system can be obtained from its differential equation. Depending on complexity of the system, degree of this equation increases. Assume that we have a system represented by mth order differential equation. In the state-space approach, the same system can be represented by m first-order differential equations. Thus, finding a solution to these in digital domain may be easier compared to the mth order differential equation. Besides, it is possible to analyze and synthesize a higher order system using state-space approach without approximating its system dynamics. Furthermore, the control system analysis using transfer function fails for a system that is initially not at rest. The state-space approach overcomes this shortcoming as well.

We only consider single input single output (SISO) systems throughout the book. Although we will not consider multi-input multi-output (MIMO) systems, we should mention one important property here. Controller design using transfer function approach becomes more complex for MIMO systems. On the other hand, the state-space approach can be directly used in MIMO system analysis and controller design. The reader should keep this in mind if such a system emerges in his or her application.

9.2 State-space Equations Representing an LTI System

In order to benefit from the state-space approach, we should first represent the system with its state-space equations. To do so, we will start with differential equations representing the LTI system in continuous-time. Then, we will switch to the discrete-time representation.

9.2.1 Continuous-time State-space Equations

Assume that a continuous-time system has the differential equation

$$\frac{d^m y(t)}{dt^m} + a_1 \frac{d^{m-1} y(t)}{dt^{m-1}} + \cdots + a_{m-1} \frac{dy(t)}{dt} + a_m y(t) = \\ b_0 \frac{d^m u(t)}{dt^m} + b_1 \frac{d^{m-1} u(t)}{dt^{m-1}} + \cdots + b_{m-1} \frac{du(t)}{dt} + b_m u(t) \tag{9.1}$$

where $u(t)$ and $y(t)$ are input and output of the system, respectively.

We can represent the generic differential equation in Eq. (9.1) by constructing the corresponding state-space representation. To do so, we should first define the states as $x_1(t), x_2(t), \ldots, x_m(t)$. Based on these, we can construct state-space equations representing the system as

$$\begin{aligned}
x_1(t) &= y(t) \\
x_2(t) &= \dot{x}_1(t) \\
&\vdots \\
x_m(t) &= \dot{x}_{m-1}(t)
\end{aligned} \tag{9.2}$$

where $\dot{x}_1(t) = dx_1(t)/dt$.

We can represent these equations in matrix form, Ogata (2009), as

$$\begin{bmatrix} \dot{x}_1 \\ \dot{x}_2 \\ \vdots \\ \dot{x}_{m-1} \\ \dot{x}_m \end{bmatrix} = \begin{bmatrix} 0 & 1 & 0 & \cdots & 0 \\ 0 & 0 & 1 & \cdots & 0 \\ \vdots & \vdots & \vdots & \cdots & \vdots \\ 0 & 0 & 0 & \cdots & 1 \\ -a_m & -a_{m-1} & -a_{m-2} & \cdots & -a_1 \end{bmatrix} \begin{bmatrix} x_1 \\ x_2 \\ \vdots \\ x_{m-1} \\ x_m \end{bmatrix} + \begin{bmatrix} 0 \\ 0 \\ \vdots \\ 0 \\ 1 \end{bmatrix} u(t) \tag{9.3}$$

$$y(t) = \begin{bmatrix} (b_m - a_m b_0) & (b_{m-1} - a_{m-1} b_0) & \cdots & (b_1 - a_1 b_0) \end{bmatrix} \begin{bmatrix} x_1 \\ x_2 \\ \vdots \\ x_{m-1} \\ x_m \end{bmatrix} + b_0 u(t) \tag{9.4}$$

We can solve Eq. (9.3) as

$$\mathbf{x}(t) = e^{\mathbf{A}(t-t_0)} \mathbf{x}(t_0) + \int_{t_0}^{t} e^{\mathbf{A}(t-\tau)} \mathbf{B} u(\tau) d\tau \tag{9.5}$$

$$\mathbf{x}(t) = \boldsymbol{\phi}(t - t_0) \mathbf{x}(t_0) + \int_{t_0}^{t} \boldsymbol{\phi}(t - \tau) \mathbf{B} u(\tau) d\tau$$

where $\mathbf{x}(t_0)$ is the initial state of the system and $e^{\mathbf{A}(t)} = \boldsymbol{\phi}(t)$ is the state-transition matrix. It is possible to find the state vector at any time t using Eq. (9.5).

Using Eqs. (9.3) and (9.4), the generic structure of a state-space model for an mth order continuous-time system can be represented as

$$\dot{x}(t) = Ax(t) + Bu(t) \tag{9.6}$$

$$y(t) = Cx(t) + Du(t) \tag{9.7}$$

where $x(t)$ is the m dimensional state vector, $u(t)$ is the input signal, and $y(t)$ is the output signal. A, B, C, and $D = b_0$ are the state (system), input, output, and feedforward matrices, respectively. For completeness, we can represent these matrices as

$$A = \begin{bmatrix} 0 & 1 & 0 & \cdots & 0 \\ 0 & 0 & 1 & \cdots & 0 \\ \vdots & \vdots & \vdots & \cdots & \vdots \\ 0 & 0 & 0 & \cdots & 1 \\ -a_m & -a_{m-1} & -a_{m-2} & \cdots & -a_1 \end{bmatrix}$$

$$B^T = \begin{bmatrix} 0 & 0 & \cdots & 0 & 1 \end{bmatrix}$$

$$C = \begin{bmatrix} (b_m - a_m b_0) & (b_{m-1} - a_{m-1} b_0) & \cdots & (b_1 - a_1 b_0) \end{bmatrix}$$

We can summarize some key observations on matrices A, B, C, and D. The A matrix describes how internal states of the system are connected to each other. In other words, it shows underlying dynamics of the system. The B matrix represents how the input signal is affecting the states. The C matrix describes how the states are combined to form system output. Finally, the D matrix is used to bypass the states and feedforward the input signal to output.

State-space equations in Eqs. (9.3) and (9.4) have a specific structure called controllable canonical form. This form becomes important when we introduce state-space controller design via pole placement in Chapter 10. There is also the observable canonical form. In this form, we will have

$$A = \begin{bmatrix} 0 & 0 & \cdots & 0 & -a_m \\ 1 & 0 & \cdots & 0 & -a_{m-1} \\ \vdots & \vdots & \vdots & \cdots & \vdots \\ 0 & 0 & \cdots & 1 & -a_1 \end{bmatrix}$$

$$B^T = \begin{bmatrix} (b_m - a_m b_0) & (b_{m-1} - a_{m-1} b_0) & \cdots & (b_1 - a_1 b_0) \end{bmatrix}$$

$$C = \begin{bmatrix} 0 & 0 & \cdots & 0 & 1 \end{bmatrix}$$

$$D = b_0$$

The controllable and observable canonical forms are the most commonly used ones. Therefore, we will revisit them in the following sections again.

9.2.2 Discrete-time State-space Equations

It is also possible to find the state vector at discrete time instants. To do so, let the sampling period be T_s and assume state values are known at time $t = nT_s$. Then, we can rewrite Eq. (9.5) as

$$\mathbf{x}((n+1)T_s) = e^{\mathbf{A}T_s}\mathbf{x}(nT_s) + \left[\int_{nT_s}^{(n+1)T_s} e^{\mathbf{A}((n+1)T_s-\tau)}\mathbf{B}\, d\tau\right] u(nT_s) \qquad (9.8)$$

The input signal $u(nT_s)$ can be taken out of integration and Eq. (9.8) can be rewritten as

$$\mathbf{x}((n+1)T_s) = \mathbf{F}\mathbf{x}(nT_s) + \mathbf{G}u(nT_s) \qquad (9.9)$$

where,

$$\mathbf{F} = e^{\mathbf{A}T_s} \qquad (9.10)$$

$$= \mathbf{I} + \mathbf{A}T_s + \frac{(\mathbf{A}T_s)^2}{2!} + \cdots$$

$$= \mathbf{I} + \mathbf{A}T_s\mathbf{\Psi}$$

where $\mathbf{\Psi} = \mathbf{I} + \mathbf{A}T_s + \frac{(\mathbf{A}T_s)^2}{2!} + \cdots$ (Ogata 1995). Here, $\mathbf{I}$ represents the identity matrix. Based on this definition, we will have

$$\mathbf{G} = \int_{nT_s}^{(n+1)T_s} e^{\mathbf{A}((n+1)T_s-\tau)}\mathbf{B}\, d\tau \qquad (9.11)$$

$$= T_s\mathbf{\Psi}\mathbf{B}$$

By focusing on sampling instants, taking $T_s = 1$, we will have

$$\mathbf{x}[n+1] = \mathbf{F}\mathbf{x}[n] + \mathbf{G}u[n] \qquad (9.12)$$

$$y[n] = \mathbf{C}\mathbf{x}[n] + \mathbf{D}u[n] \qquad (9.13)$$

As a result, we obtain the discrete-time state-space representation starting from continuous-time differential equation. We can follow a more straightforward methodology as well. To do so, we will consider conversion between state-space and transfer function representations in Section 9.3.

9.2.3 Representing Discrete-time State-space Equations in Code Form

As can be seen in Eqs. (9.12) and (9.13), state-space representation of an LTI system only requires defining four matrices as $\mathbf{F}$, $\mathbf{G}$, $\mathbf{C}$, and $\mathbf{D}$. We can benefit from the matrix operations in C or Python languages, as explained in Chapter 3, to realize the state-space representation. Although this is a valid option, we will benefit from the Python and MicroPython control systems library structures and functions in this book. These will simplify life for us.

Before going further, let us consider the LTI system represented by its state-space form as

$$\mathbf{x}[n+1] = \begin{bmatrix} 1.8188 & -0.8224 \\ 1.0 & 0.0 \end{bmatrix} \mathbf{x}[n] + \begin{bmatrix} 1.0 \\ 0.0 \end{bmatrix} u[n] \tag{9.14}$$

$$y[n] = \begin{bmatrix} 0.0360 & 0.0017 \end{bmatrix} \mathbf{x}[n] + 0.0094u[n] \tag{9.15}$$

We can represent these matrices in Python as $F = [[1.8188, -0.8224], [1.0, 0.0]]$, $G = [[1.0], [0.0]]$, $C = [[0.036, 0.0017]]$, and $D = [[0.0094]]$. We can benefit from MicroPython control systems library ss structure to represent them. We provide such an example in Listing 9.1.

Listing 9.1: State-space representation example.

```
import mpcontrol
import mpcontrolPC

Ts = 0.0005

F = [[1.8188, -0.8224], [1.0, 0.0]]
G = [[1.0], [0.0]]
C = [[0.036, 0.0017]]
D = [[0.0094]]

ss = mpcontrolPC.ss(F, G, C, D, Ts)

N = 1000

u = mpcontrol.step_signal(N, 1)

y = mpcontrol.lsim_ss(ss, u, N, 0)

x1 = mpcontrol.lsim_ss(ss, u, N, 1)
x2 = mpcontrol.lsim_ss(ss, u, N, 2)
```

We can benefit from MicroPython control systems library to simulate the system represented in state-space form. To do so, we should use the function $lsim\_ss$ in the library. This function is constructed such that it accepts initial states. It also feeds state values in time when desired. We provide such an example in Listing 9.1. Here, we feed a unit-step signal to the system and obtain output y as well as state values $x1$ and $x2$. We plot these values in Figure 9.1.

Let us look at Figure 9.1 more closely. As can be seen in Figure 9.1(a), unit-step response of the system reaches a final value. Two states of the system also reach their final value as the time index increases as can be seen in Figures 9.1(b) and 9.1(c). Finally, we can plot the state-space as in Figure 9.1(d). This figure shows how states of the system move in time on the state-space. Since the initial state

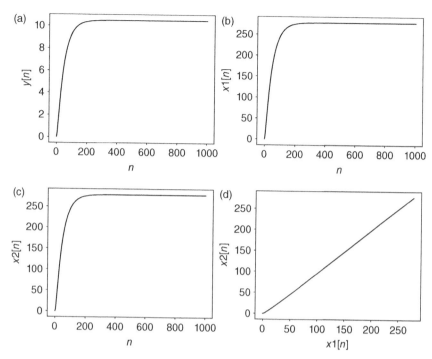

Figure 9.1 State-space representation of the selected system. (a) Output. (b) First state. (c) Second state. (d) State space.

values were taken as zero for both states, both states started with this value and reached a final point in state-space.

9.3 Conversion Between State-space and Transfer Function Representations

State-space and transfer function representations both depend on difference equation of the system. Therefore, it is possible to convert one form to another. We will focus on these conversions in this section.

9.3.1 From Transfer Function to State-space Equations

We can use the Python and MicroPython control systems libraries together to obtain state-space equations of a system from its transfer function. To do so, we can use the function `tf2ss` within the Python control systems library. Then, we can convert this representation to the MicroPython form using the function `ss_to_mpss`. We provide usage of these functions starting from the continuous-time DC motor transfer function in Listing 9.2. Here, we start by defining the continuous-time transfer function and finally obtain its discrete-time state-space representation. We can use this representation for further operations.

Listing 9.2: Obtaining state-space representation of the DC motor from its transfer function.

```
import mpcontrolPC
import control

num = [165702.47]
den = [1, 390.2, 15701]

tfs = mpcontrolPC.tf(num, den)

Ts = 0.0005
tfz = mpcontrolPC.c2d(tfs, Ts)

tf_cl = mpcontrolPC.mptf_to_tf(tfz)
ss_cl = control.tf2ss(tf_cl)

ss = mpcontrolPC.ss_to_mpss(ss_cl)
```

We can also benefit from the Python control systems library to obtain the controllable and observable state-space representations of a system. To do so, we should use the functions `canonical_form()` and `observable_form()` within the library. Let us apply these functions to the state-space representation obtained in Listing 9.2. We provide the corresponding Python code in Listing 9.3. We can benefit from the representation suitable for our usage.

Listing 9.3: Obtaining the canonical forms of state-space representation of the DC motor.

```
import mpcontrolPC
import control

num = [165702.47]
den = [1, 390.2, 15701]

tfs = control.tf(num, den)

Ts = 0.0005
tfz = control.c2d(tfs, Ts)

ss = control.tf2ss(tfz)

ss_mp = mpcontrolPC.ss_to_mpss(ss)
print('Original form')
print(ss_mp)

ss_cf,T_cf = control.canonical_form(ss,'reachable')
print('Reachable form')
print(ss_cf)

ss_of,T_of = control.observable_form(ss)
print('Observable form')
print(ss_of)
```

As we execute the Python code Listing 9.3, we will obtain the controllable and observable canonical forms and find out if this state-space representations are controllable and observable. The controllable canonical form for the DC motor becomes

$$\mathbf{x}[n+1] = \begin{bmatrix} 1.8192 & -0.8228 \\ 1.0 & 0.0 \end{bmatrix} \mathbf{x}[n] + \begin{bmatrix} 1.0 \\ 0.0 \end{bmatrix} u[n] \tag{9.16}$$

$$y[n] = \begin{bmatrix} 0.0194 & 0.0182 \end{bmatrix} \mathbf{x}[n] \tag{9.17}$$

Likewise, the observable canonical form for the DC motor becomes

$$\mathbf{x}[n+1] = \begin{bmatrix} 1.8192 & 1.0 \\ -0.8228 & 0.0 \end{bmatrix} \mathbf{x}[n] + \begin{bmatrix} 0.0194 \\ 0.0182 \end{bmatrix} u[n] \tag{9.18}$$

$$y[n] = \begin{bmatrix} 1.0 & 0.0 \end{bmatrix} \mathbf{x}[n] \tag{9.19}$$

As can be seen here, there is a duality between controllable and observable canonical forms. We can see that the system matrix $\mathbf{F}$ in the observable canonical form is transpose of the system matrix in the controllable canonical form. Similarly, the input matrix $\mathbf{G}$ in observable canonical form is transpose of output matrix $\mathbf{C}$ in the controllable canonical form. Furthermore, the output matrix $\mathbf{C}$ in the observable canonical form is transpose of the input matrix $\mathbf{G}$ in the controllable canonical form. This duality property can be used to reduce the amount of work needed to prove controllability and observability properties of a system, to be introduced in Section 9.4. It will also be useful in designing controllers and observers to be explained in Chapter 10.

9.3.2 From State-space Equations to Transfer Function

It is possible to find the transfer function of a system when its state-space equations are given. Taking z-transform on both sides in Eq. (9.12), we obtain

$$z\mathbf{X}(z) - z\mathbf{x}[0] = \mathbf{F}\mathbf{X}(z) + \mathbf{G}U(z) \tag{9.20}$$

$$Y(z) = \mathbf{C}\mathbf{X}(z) + \mathbf{D}U(z) \tag{9.21}$$

where $\mathbf{x}[0]$ is the initial state of the system. To note here, initial states are assumed to be zero to find the transfer function of a system. The system transfer function is the ratio of output, $Y(z)$, to input, $U(z)$ as mentioned in Section 4.5. Therefore, transfer function of the system represented by its state-space equations becomes

$$G(z) = \frac{Y(z)}{U(z)} = \mathbf{C}(z\mathbf{I} - \mathbf{F})^{-1}\mathbf{G} + \mathbf{D} \tag{9.22}$$

We can also use the Python control systems library to find transfer function of the system if its state-space equations are known. To do so, we can use the function `ss2tf()` within the library. Let us take state-space representation of the DC motor given in Eqs. (9.14) and (9.15). We can use the Python code given in Listing 9.4 to obtain transfer function of the DC motor.

Listing 9.4: Obtaining transfer function of the DC motor from its state-space equations.

```
import control

#System Description in State-Space
F = [[1.8188, -0.8224], [1,0]]
G =   [[1], [0]]
C = [[0.0360, 0.0017]]
D = [[0.0094]]

Ts=0.0005

#Transfer Function of the System
tf_system = control.ss2tf(F,G,C,D,Ts)

print (tf_system)
```

As we execute the Python code Listing 9.4, we obtain transfer function of the DC motor as

$$G(z) = \frac{0.0094z^2 + 0.0189z + 0.0094}{z^2 - 1.8190z + 0.8224} \tag{9.23}$$

This is in fact the original transfer function of the DC motor given in Eq. (6.18). Therefore, we were able to reach it from another way.

9.4 Properties of the System from its State-space Representation

We can obtain properties of the LTI system from its state-space representation as well. To do so, we will introduce the time-domain analysis first. Then, we will check the stability of the system. Afterward, we will focus on two important system properties specific to the state-space representation as controllability and observability. Controllability determines the effectiveness of state feedback control. Observability determines the possibility of state estimation from input and output measurements.

9.4.1 Time Domain Analysis

We can use the Python control systems library to obtain ζ and ω_n of the system directly from its state-space equations. To do so, we can use the function damp() within the library. To note here, these two values are calculated from the state-space representation.

Let us take state-space representation of the DC motor obtained from its continuous-time transfer function in Listing 9.2. We can extend this code to obtain ζ and ω_n. To do so, we form the extended Python code as in Listing 9.5. As we execute the Python code Listing 9.5, we will obtain ζ and ω_n values of the DC motor same as in Section 7.1.1.

Listing 9.5: Obtaining ζ and ω_n of the DC motor from its state-space representation.

```
import control

num = [165702.47]
den = [1, 390.2, 15701]

tfs = control.tf(num, den)

Ts = 0.0005
tfz = control.c2d(tfs, Ts)

ss = control.tf2ss(tfz)

wn,zeta,poles=control.damp(ss)

print('wn', wn)
print('zeta', wn)
print('poles', poles)
```

9.4.2 Stability

We can check stability of a system from its state-space representation using two different methods. We can use the `pole()` function within the Python control systems library as the first and straightforward method. Then, we can check whether any system pole is outside the unit-circle or not.

Let us take the state-space representation of the DC motor obtained from its continuous-time transfer function in Listing 9.2. We can extend this code to obtain poles of the system directly from its state-space representation. To do so, we form the extended Python code as in Listing 9.6.

Listing 9.6: Obtaining poles of the DC motor from its state-space representation.

```
import control

num = [165702.47]
den = [1, 390.2, 15701]

tfs = control.tf(num, den)

Ts = 0.0005
tfz = control.c2d(tfs, Ts)

ss = control.tf2ss(tfz)

poles=control.pole(ss)

print(poles)

wn,zeta,poles=control.damp(ss)
```

As we execute the Python code Listing 9.6, we obtain poles of the DC motor as 0.9775 and 0.8417. These values are the same as in Chapter 6. Besides, they are in the unit circle. Hence, we can deduce that the DC motor is stable.

As the second method, we can benefit from the state matrix **F**. Poles of the system can be obtained using the equation $|z\mathbf{I} - \mathbf{F}| = 0$ (for $\mathbf{D} = 0$, no direct feedforward path). Hence, roots of this equation are in fact eigenvalues of the matrix **F**. We can use the numpy library under Python for this purpose.

Let us take state-space representation of the DC motor obtained from its continuous-time transfer function in Listing 9.2. We can extend this code to obtain poles of the system directly from its state-space representation. To do so, we form the extended Python code as in Listing 9.7. As we execute the code, we obtain the eigenvalues as 0.9775 and 0.8417. These are the actual poles of the system. Besides, they are within the unit circle. Again, we can deduce that the DC motor is stable.

Listing 9.7: Obtaining poles of the DC motor from the eigenvalues of the **F** matrix.

```
import control
import numpy as np

num = [165702.47]
den = [1, 390.2, 15701]

tfs = control.tf(num, den)

Ts = 0.0005
tfz = control.c2d(tfs, Ts)

ss = control.tf2ss(tfz)

eig_F,eigv_F = np.linalg.eig(ss.A)

print(eig_F)
```

9.4.3 Controllability

A discrete-time system is said to be completely controllable if an input sequence $u[n]$ can be determined that will take the system to any desired state $x[n]$ in finite number of time steps. Here, we assume that the system model and its state $x[n]$ are known at the specific time step n. Controllability is checked by looking at rank of the matrix $\begin{bmatrix} \mathbf{G} & \mathbf{FG} & \mathbf{F}^2\mathbf{G} & \cdots & \mathbf{F}^{m-1}\mathbf{G} \end{bmatrix}$. If the rank of this matrix is not equal to zero, then the system is controllable.

We can benefit from the Python control systems library to check controllability using state-space equations of the system. To do so, we can use the function ctrb() within the library to find the controllability matrix. The controllability property of the DC motor can be checked using the Python code given in Listing 9.8 on PC. As we execute this code, we obtain the controllability matrix and its rank. Since rank of the controllability matrix is two, we can deduce that the DC motor is controllable.

Listing 9.8: Checking controllability of the DC motor in Python.

```
import control
import numpy as np

F = [[1.8188, -0.8224], [1.0, 0.0]]
G = [[1.0], [0.0]]
C = [[0.036, 0.0017]]
D = [[0.0094]]

Cont_matrix =control.ctrb(F, G)
print(Cont_matrix)

Rank_Cont_matrix = np.linalg.matrix_rank(Cont_matrix)
print(Rank_Cont_matrix)
```

9.4.4 Observability

A discrete-time system is said to be completely observable if its state $x[n]$ at any specific step n can be determined from the system model, input, and output values for a finite number of steps. Observability refers to the ability to estimate a state variable. Observability is checked by looking at the rank of the matrix $\begin{bmatrix} C & CF & CF^2 & \cdots & CF^{m-1} \end{bmatrix}^T$. If the rank of this matrix is not equal to zero, then the system is observable.

We can benefit from the Python control systems library to find the observability using state-space equations of the system. To do so, we can use the function obsv() within the library to find the observability matrix. The observability property of the DC motor can be checked using the Python code given in Listing 9.9 on PC. As we execute this code, we obtain the observability matrix and its rank. Since rank of the observability matrix is two, we can deduce that the DC motor is observable.

Listing 9.9: Checking observability of the DC motor in Python.

```
import control
import numpy as np

F = [[1.8188, -0.8224], [1.0, 0.0]]
G = [[1.0], [0.0]]
C = [[0.036, 0.0017]]
D = [[0.0094]]

Obs_matrix =control.obsv(F, C)
print(Obs_matrix)

Rank_Obs_matrix = np.linalg.matrix_rank(Obs_matrix)
print(Rank_Obs_matrix)
```

The controllability and observability concepts are introduced by Kalman (1960, 1963). They will be useful when designing state feedback controller and state estimator to be explained in state-space method based controller design in Chapter 10. There, we will see that if the system is controllable and observable, then we can accomplish the design objective of placing the poles precisely at desired locations to meet performance criteria (such as less e_{ss} and T_s).

We can also relate controllability and observability with the transfer function of a system. If transfer function of the system does not have pole-zero cancellation (some poles and zeros of the transfer function are same and they are cancelled), then the system can be represented in controllable and observable forms. If transfer function of the system has pole-zero cancellation, then the system will be either uncontrollable or unobservable.

9.5 Application: Observing States of the DC Motor in Time

In this application, we will simulate state-space representation of the DC motor on the STM32 microcontroller. The aim here is twofold. First, we will provide a method to implement the state-space representation in C and Python languages on the STM32 microcontroller. Second, we will form a setup to be used for state-space based controller implementation in Chapter 10.

9.5.1 Hardware Setup

We will observe the simulation results on PC. Therefore, the reader only needs the FT232 module and its connections on the STM32 microcontroller. Please see Section 4.6 for more detail on this setup.

9.5.2 Procedure

We will use state-space representation of the DC motor introduced in Section 9.3.1 within this application. Afterward, a step input with 8 V amplitude and 500 elements will be applied to the system and simulation results will be obtained. We will transfer state values in time and output of the system to PC as we did in Section 4.6. As the final part of the application, we will plot the obtained results on PC via Python.

9.5.3 C Code for the System

We should form a new project in Mbed Studio and use the C code in Listing 9.10 for the application. For the software setup, the reader should add `Ccontrol.cpp` and `Ccontrol.h` files to the project. As we execute this code, state-space system is formed within the microcontroller using the function `create_ss(ss_struct *ss, float *F, float *G, float *C,float *D, int no_states, float ts)`. Here, `ss` is the created state space structure. F, G, C, and D are state space matrices. `no_states` is the state number and `ts` is the sampling period for the system. We feed the step input signal to the system and obtain its output and state variables using the function `lsim_ss(ss_struct *ss, float *u, float *x0, float *result, int select, int N)`. Here, `ss` is the state space structure. `u` is the input array. `x0` is the initial states array. `result` is the simulation output array. `select` is the simulation output selection (0 for system output, 1 for the first state variable, 2 for the second state variable, and so on). N is the number of elements obtained from simulation.

Listing 9.10: The C code to be used in simulating the state-space system on the STM32 microcontroller.

```
#include "mbed.h"
#include "Ccontrol.h"

#define no_states 2
#define N 500

RawSerial device(PD_5, PD_6, 921600);

ss_struct ss_system;
float input[N], output[N], state1[N], state2[N];
float Ts = 0.0005;

float F_ss[no_states][no_states] = {
    {1.8188, -0.8224},
    {1.0, 0.0}
};

float G_ss[no_states][1] = {
    {1.0},
    {0.0}
};

float C_ss[1][no_states] = {
    {0.036011284, 0.001674768}
};

float D_ss[1][1] = {
    {0.00943}
};

float x0_ss[no_states][1] = {
    {0.0},
    {0.0}
};

int main()
{
    create_ss(&ss_system, &F_ss[0][0], &G_ss[0][0], &C_ss[0][0], &
```

```
        D_ss[0][0], no_states, Ts);
    step_signal(N, 8, input);
    lsim_ss(&ss_system, input, &x0_ss[0][0], output, 0, N);
    lsim_ss(&ss_system, input, &x0_ss[0][0], state1, 1, N);
    lsim_ss(&ss_system, input, &x0_ss[0][0], state2, 2, N);
    while(true) {
        while(device.getc() != 'r');
        send_data(output,N);

        while(device.getc() != 'r');
        send_data(state1,N);

        while(device.getc() != 'r');
        send_data(state2,N);
    }
}
```

In Listing 9.10, we send output and state variables to PC for further processing. On the PC side, we benefit from the functions introduced in previous applications. We provide a Python code for this purpose in Listing 9.11. Here, we use the function `get_data_from_MC(NoElements, com_port, selection)` three times to receive output and state arrays separately. As we receive them, we can save them to separate files. Afterward, they can be plotted on PC if desired.

Listing 9.11: The Python code to receive and store state values on PC.

```
import mpcontrolPC
import matplotlib.pyplot as plt

N = 500
output_signal = mpcontrolPC.get_data_from_MC(N, 'com10', 0)
state1 = mpcontrolPC.get_data_from_MC(N, 'com10', 0)
state2 = mpcontrolPC.get_data_from_MC(N, 'com10', 0)

mpcontrolPC.save_to_file(output_signal, file_name='output.dat')
mpcontrolPC.save_to_file(state1, file_name='state1.dat')
mpcontrolPC.save_to_file(state2, file_name='state2.dat')
```

9.5.4 Python Code for the System

We can repeat this application in MicroPython on the STM32 microcontroller. The reader can use the Python code in Listing 9.12 for this purpose. Obtaining and saving data, and plotting results on PC are the same as in Section 9.5.3. Therefore, please consult that section for implementation details.

Listing 9.12: The Python code to be used in simulating the state-space system on the STM32 microcontroller.

```
import pyb
import mpcontrol
import struct

uart2 = pyb.UART(2, 921600)
```

```
N = 500

def send_data(array, length):
        for i in range(length):
                dummy_input = int(struct.unpack('<I', struct.pack
                        ('<f', array[i]))[0])
                dummy_send = dummy_input & 0x000000ff
                uart2.writechar(dummy_send)
                dummy_send = (dummy_input & 0x0000ff00) >> 8
                uart2.writechar(dummy_send)
                dummy_send = (dummy_input & 0x00ff0000) >> 16
                uart2.writechar(dummy_send)
                dummy_send = (dummy_input & 0xff000000) >> 24
                uart2.writechar(dummy_send)

def main():
        F = [[1.8188, -0.8224], [1.0, 0.0]]
        G = [[1.0], [0.0]]
        C = [[0.036011284, 0.001674768]]
        D = [[0.00943]]
        Ts = 0.0005
        ss_system = mpcontrol.ss(F, G, C, D, Ts)
        input = mpcontrol.step_signal(N, 8)

        output = mpcontrol.lsim_ss(ss_system, input, N, 0, [0, 0])
#       state1 = mpcontrol.lsim_ss(ss_system, input, N, 1, [0, 0])
#       state2 = mpcontrol.lsim_ss(ss_system, input, N, 2, [0, 0])

        while True:
                while uart2.readchar() != 0x72:
                        pass
                send_data(output,N)
#               send_data(state1,N)
#               send_data(state2,N)

main()
```

9.5.5   Observing Outputs

We provide the obtained results in Figure 9.2 when the C code of the application is executed. The same results can be obtained when the Python code is executed. Therefore, we did not repeat the procedure again.

9.6   Summary

We focused on discrete system analysis by state-space representation in this chapter. To do so, we started with the definition of state and how it can be represented in code form. Afterward, we explored how transfer function of an LTI system can be represented in state-space form. We also considered how a state-space representation can be transformed to transfer function form. We also explored system properties based on its state-space representation. We finally provided an end of chapter application based on C and Python. Although this

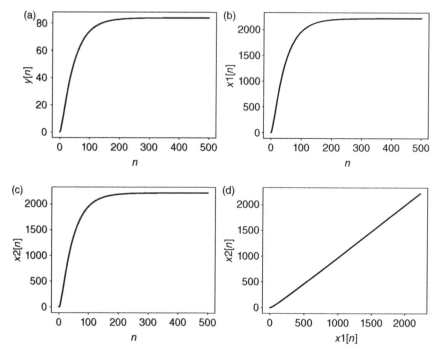

Figure 9.2 State-space and output plots of the DC motor obtained by the C code. (a) Output. (b) First state. (c) Second state. (d) State space.

application is a simulation running on the STM32 microcontroller, it forms the setup for the state-space system control to be introduced in the next chapter.

Problems

9.1 Consider a system represented by

$$x[n+1] = \begin{bmatrix} 0.9990 & 0.0095 \\ -0.1903 & 0.9039 \end{bmatrix} x[n] + \begin{bmatrix} 0.0 \\ 0.0095 \end{bmatrix} u[n] \tag{9.24}$$

$$y[n] = \begin{bmatrix} 1 & 0 \end{bmatrix} x[n] \tag{9.25}$$

Plot unit-step response and states of the system when $T_s = 0.0005$ seconds.

9.2 Consider a DC motor with the transfer function

$$G(s) = \frac{5}{s^2 + 20s} \tag{9.26}$$

a. convert this system to digital form by bilinear transformation with $T_s = 0.0005$ seconds.
b. find state-space equations of the digital system.
c. plot the response and states of the digital system when a unit pulse signal is applied to it.

9.3 Consider the digital system in Problem 9.2,
a. find its controllable and observable forms of state-space equations of the system.
b. plot the unit-step response for the controllable and observable forms of the system.

9.4 Find transfer function of the system given in Problem 9.1.

9.5 Find the poles, ω_n, and ζ of the system given in Problem 9.4.

9.6 Find the poles, ω_n, and ζ of the transfer function given in Problem 9.2.

9.7 Find the controllability matrix and its rank for the system given in Problem 9.1.

9.8 Find the observability matrix and its rank for the system given in Problem 9.1.

9.9 Consider the RC filter with continuous-time transfer function

$$G(s) = \frac{1000}{s + 1000} \tag{9.27}$$

a. convert this system to digital form by bilinear transformation with $T_s = 0.0005$ seconds.
b. find the state-space equations of the digital system.
c. plot the response and states of the digital system when the unit step signal is applied to it.
d. find controllable and observable forms of state-space equations for the digital system.
e. plot the response of the system for the unit-step input when the controllable and observable forms of the digital system are used.
f. find the controllability matrix and its rank for the system.
g. find the observability matrix and its rank for the system.

9.10 Consider the digital system given in Problem 9.9
  a. plot the response and states of the system when a sinusoidal signal is applied to it.
  b. plot the response and states of the system when a rectangular signal is applied to it.
  c. plot the response and states of the system when a random signal is applied to it.

10

State-space Based Controller Design

We introduced transfer function based controller design in Chapter 8. The idea there was using time and frequency domain properties of the LTI system in the controller design. State-space representation introduced in Chapter 9 provides other options in designing controllers. We will consider one such option, called pole placement, in this chapter. Before introducing this method, we should understand the general layout for state-space based controller design. Therefore, we will start with explaining the necessary information to be used throughout the chapter. Then, we will provide mathematical background for the controller design based on pole placement. Afterward, we will show how such controllers can be designed and implemented in code form. A state-space based controller may depend on estimating state values from the input and output signals. Therefore, we will introduce mathematical background for the observer structure to be used for this purpose. Then, we will provide the observer-based controller implementation in code form. Finally, we will provide an end of chapter application showing how the DC motor can be controlled via state-space based methods both in C and Python languages.

10.1 General Layout

The state-space based controller design is different from the previously introduced transfer function based methods in Chapter 8. Besides, there are several phases in the design. Therefore, it becomes a necessity to briefly introduce them first. This section serves this purpose.

Embedded Digital Control with Microcontrollers: Implementation with C and Python,
First Edition. Cem Ünsalan, Duygun E. Barkana, and H. Deniz Gürhan.
© 2021 The Institute of Electrical and Electronics Engineers, Inc. Published 2021 by John Wiley & Sons, Inc.
Companion website: www.wiley.com/go/Unsalan/Embedded_Digital_Control_with_Microcontrollers

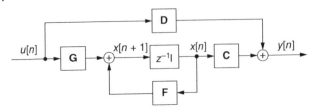

Figure 10.1 State-space form of an LTI system.

10.1.1 Control Based on State Values

States representing the system may be used in forming a controller. As explained in Chapter 9, we can represent an LTI system in state-space form as

$$\mathbf{x}[n+1] = \mathbf{Fx}[n] + \mathbf{Gu}[n]$$
$$y[n] = \mathbf{Cx}[n] + \mathbf{Du}[n]$$

We can visualize this state-space form as in Figure 10.1. To note here, all operations in this figure are matrix based.

In order to form a controller to the LTI system represented in Figure 10.1, it should be controllable as explained in Chapter 9. Hence, we can obtain the desired response for a given input signal. If the system is uncontrollable, all is not lost. We can still obtain an acceptable response by designing a controller. Here, we will only benefit from the controllable states in the design and ask for the remaining states to be at least providing a stable response. More information on this topic can be found in the book by Ogata (2009).

We will limit ourselves by only controllable systems throughout this book. Hence, we can represent the state-space form in Figure 10.1 by its controllable canonical form as

$$\mathbf{x}[n+1] = \mathbf{Fx}[n] + \mathbf{Gu}[n]$$
$$y[n] = \mathbf{Cx}[n]$$

Then, this state-space form can be represented as in Figure 10.2. We will base our control actions from the representation in this figure. Here, we have two options as regulator and controller structures, to be explored next.

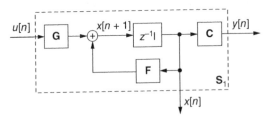

Figure 10.2 The controllable canonical state-space form of an LTI system.

10.1.2 Regulator Structure

Controllers designed till this point (especially closed-loop controllers) had a reference input to be followed. Performance of the controller was judged by its reference following action. The standard approach in state-space based design starts by setting the reference input to zero. This specific structure is called regulator. Here, the aim is forming a controller such that the system responds to any disturbance by turning back to zero state all the times. Hence, the system becomes more robust against disturbances. In other words, if a disturbance affects the system, the regulator leads it to its zero state after some time.

We provide how the regulator structure can be formed by the state-space representation of a system in Figure 10.3. Here, state values, weighted by constant gain, are fed back to input of the system as a control signal. We will show how such a regulator can be constructed in Section 10.2.

10.1.3 Controller Structure

The regulator structure may not be sufficient for most control actions in which the system should follow a reference input. To do so, the regulator structure should be modified such that a reference signal can be fed to it. This results in the controller structure as in Figure 10.4.

In Figure 10.4, we still have the feedback loop formed by the weighted state values. However, we also have an extra feedback loop and form an error signal based on the reference and output signals. Besides, we modify this error signal by an integrator before feeding it to the system. We will explain why these steps are necessary to form a controller structure in Section 10.2.

Figure 10.3 The regulator structure.

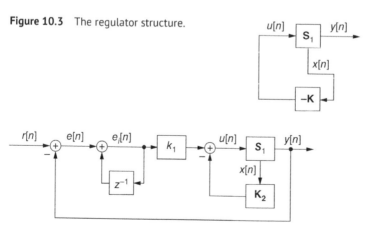

Figure 10.4 The controller structure.

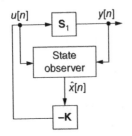

Figure 10.5 Observer-based regulator structure.

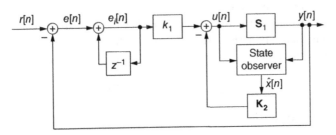

Figure 10.6 Observer-based controller structure.

10.1.4 What if States Cannot be Measured Directly?

The regulator and controller structures depend on state values for generating the control signal. There may be cases in which these state values cannot be measured directly in practice. Hence, the control signal cannot be generated for the regulator and controller structures. We can use an observer to overcome this problem. To note here, we should have an observable system at hand for this purpose.

The observer takes the system output with control input and generates the estimated state values based on these. Then, the estimates can be used in the control action. We provide the observer-based regulator structure in Figure 10.5. Likewise, we provide the observer-based controller structure in Figure 10.6. We will explain the design procedure for both structures in Section 10.4.

10.2 Regulator and Controller Design via Pole Placement

Pole placement is a fairly straightforward method in designing the regulator and controller in state-space. Therefore, we will introduce it in this section starting from the fundamentals. Then, we will explain how the design can be specialized for the regulator and controller.

10.2.1 Pole Placement

As explained in the previous section, state-space based regulator and controller structures benefit from the feedback loop from state variables. The pole placement method aims to set closed-loop poles of the system by adjusting the state feedback gain, $\mathbf{K}$, values. Here, roots of the closed-loop system are placed where the performance requirements, especially in time-domain, are satisfied. Let us consider this method for the regulator and controller structures next.

10.2.2 Regulator Design

Consider the state-space model of an LTI system

$$\mathbf{x}[n+1] = \mathbf{Fx}[n] + \mathbf{G}u[n]$$
$$y[n] = \mathbf{Cx}[n]$$

We can form the regulator structure by forming the input signal as

$$u[n] = -\mathbf{Kx}[n] \tag{10.1}$$

where $\mathbf{K}$ is the $1 \times m$ state feedback gain matrix.

When the input signal in Eq. (10.1) is substituted to the state-space model of the LTI system, it becomes

$$\mathbf{x}[n+1] = (\mathbf{F} - \mathbf{GK})\mathbf{x}[n] \tag{10.2}$$
$$y[n] = \mathbf{Cx}[n] \tag{10.3}$$

We can take the z-transform of Eqs. (10.2) and (10.3) and form the denominator polynomial of the (closed-loop) system as $|z\mathbf{I} - \mathbf{F} + \mathbf{GK}| = 0$. We know that poles of the closed-loop system are determined by the roots of this equation. Assume that, we want to set the roots of the equation as $\{\lambda_1, \ldots, \lambda_m\}$. The pole placement method determines the state feedback (regulator) gain $\mathbf{K}$ based on these values. To do so, the straightforward method is using Ackermann's formula, to be explored next.

10.2.3 Ackermann's Formula for the Regulator Gain

Without loss of generality, let us consider the system to be controlled as second-order. It is possible to define poles of the desired system as $(z - \lambda_1)(z - \lambda_2)$. Here, λ_1 and λ_2 values can be set to satisfy the time domain requirements. Then, we can equate the desired and actual denominator polynomials to obtain the $\mathbf{K}$ matrix entries. Although this method works for the second-order system, it is difficult to extend this approach to higher order systems.

Ackermann's formula can be used to obtain the entries of **K** for any order system (Ogata 1995). The method works as follows. First, we should decide on the desired poles for the mth order system. Then, the desired denominator polynomial for this system can be formed as $\alpha_r(z) = (z - \lambda_1)(z - \lambda_2) \cdots (z - \lambda_m) = 0$ as mentioned before. Based on this representation, Ackermann's formula for the gain matrix in the regulator structure becomes

$$\mathbf{K} = \begin{bmatrix} 0 & 0 & \cdots & 1 \end{bmatrix} \begin{bmatrix} \mathbf{G} & \mathbf{FG} & \mathbf{F}^2\mathbf{G} & \cdots & \mathbf{F}^{m-1}\mathbf{G} \end{bmatrix}^{-1} \alpha_r(\mathbf{F}) \tag{10.4}$$

The obtained **K** matrix entries can be directly used in forming the state feedback for the regulator structure.

10.2.4 Controller Design

The controller design based on the pole placement methodology follows the same steps as in the regulator design. However, the state-space representation is modified based on Figure 10.4. As can be seen in this figure, there are two set of gain values to be calculated as k_1 and $\mathbf{K}_2$. Besides, an integrator is added before feeding the error signal to the system. Hence, the signal $e[n]$ is modified to obtain $e_i[n]$. Such a modification is necessary if the system to be controlled does not have an integrator within itself.

In order to form the controller, we should obtain the gain values k_1 and $\mathbf{K}_2$. To do so, the state-space representation of the system is modified such that the error signal is also added as a new state value. Here, we will not provide the mathematical derivations for this setup. For a more detailed analysis and derivation please see the book by Ogata (2009). Based on the modified state-space representation for the controller structure, we will have

$$\mathbf{x}[n + 1] = \hat{\mathbf{F}}\mathbf{x}[n] + \hat{\mathbf{G}}u[n] \tag{10.5}$$

$$y[n] = \hat{\mathbf{C}}\mathbf{x}[n] \tag{10.6}$$

where

$$u[n] = k_1 e_i[n] - \mathbf{K}_2\mathbf{x}[n] \tag{10.7}$$
$$= k_1(e[n] + e_i[n-1]) - \mathbf{K}_2\mathbf{x}[n]$$
$$= k_1(r[n] - y[n-1] + e_i[n-1]) - \mathbf{K}_2\mathbf{x}[n]$$

Based on the derivation in Ogata (2009), we can form the modified matrices in Eqs. (10.5) and (10.6) as

$$\hat{\mathbf{F}} = \begin{bmatrix} \mathbf{F} & 0 \\ -\mathbf{C} & 0 \end{bmatrix} \tag{10.8}$$

$$\hat{\mathbf{G}}^T = \begin{bmatrix} \mathbf{G}^T & 0 \end{bmatrix} \tag{10.9}$$

$$\hat{\mathbf{C}} = \begin{bmatrix} \mathbf{C} & 0 \end{bmatrix} \tag{10.10}$$

As can be seen in Eqs. (10.5) and (10.6), the modified controller representation has the structure similar to the regulator case. Here, we can form the gain matrix for the controller as

$$\hat{\mathbf{K}} = \begin{bmatrix} \mathbf{K}_2 & k_1 \end{bmatrix} \tag{10.11}$$

Following the regulator design given in the previous section, poles of the closed-loop system with the state-feedback is obtained by the denominator polynomial as $|z\mathbf{I} - \hat{\mathbf{F}} + \hat{\mathbf{G}}\hat{\mathbf{K}}| = 0$. Here, we should also add another pole to the desired denominator polynomial to handle the new state. Hence, the new set of desired poles become $\{\lambda_1, \ldots, \lambda_m, \lambda_{m+1}\}$. For most cases, we can set λ_{m+1} to a value close to zero. The pole placement method determines the state feedback gain $\hat{\mathbf{K}}$ based on these values. Afterward, we can extract $\mathbf{K}_2$ and k_1 using Eq. (10.11).

10.2.5 Ackermann's Formula for the Controller Gain

Ackermann's formula can also be used to obtain the gain matrix in Eq. (10.11) for the controller structure. Based on the new state definitions in this structure, the desired denominator polynomial can be constructed as $\alpha_c(z) = (z - \lambda_1)(z - \lambda_2) \cdots (z - \lambda_m)(z - \lambda_{m+1}) = 0$. Now, Ackermann's formula for the gain matrix in Eq. (10.11) becomes

$$\hat{\mathbf{K}} = \begin{bmatrix} 0 & 0 & \cdots & 1 \end{bmatrix} \begin{bmatrix} \hat{\mathbf{G}} & \hat{\mathbf{F}}\hat{\mathbf{G}} & \hat{\mathbf{F}}^2\hat{\mathbf{G}} & \cdots & \hat{\mathbf{F}}^m\hat{\mathbf{G}} \end{bmatrix}^{-1} \alpha_c(\hat{\mathbf{F}}) \tag{10.12}$$

As we obtain $\hat{\mathbf{K}}$ from Eq. (10.12), we can extract $\mathbf{K}_2$ and k_1 from Eq. (10.11).

10.3 Regulator and Controller Design in Python

Python can be used to design the regulator and controller structures. Here, the Python control systems library will be of great help. Let us start with the regulator design next.

10.3.1 Regulator Design

Python control systems library has two built-in functions as `acker()` and `place()` to calculate the regulator gain matrix **K** from state-space representation of the system. More information on these functions can be obtained from the library website. The `acker()` function has three inputs as F, G, and `poles`. These correspond to the matrices **K**, **G**, and desired pole locations. Hence, we should call the function as `acker(F, G, poles)`. Let us provide a sample application on the usage of this function.

Example 10.1 *Finding the regulator gain matrix for the DC motor*

We pick the DC motor as an example. Therefore, we use its controllable state space representation. For our example, we set the desired closed loop poles for the pole placement method as $[0.9444+0.1134j, 0.9444-0.1134j]$. Note that selection of the desired closed-loop poles is important since their location corresponds to eigenvalues of the system. These set the system response characteristics. We can use the Python code in Listing 10.1 on PC to calculate the regulator gain matrix. As we execute the code, we obtain $\mathbf{K} = \begin{bmatrix} -0.0697 & 0.0821 \end{bmatrix}$.

Listing 10.1: Gain matrix for the regulator.

```
import control

num = [165702.47]
den = [1, 390.2, 15701]

tfs = control.tf(num, den)

Ts = 0.0005
tfz = control.c2d(tfs, Ts)

ss = control.tf2ss(tfz)

ss_cf,T_cf = control.canonical_form(ss,'reachable')

poles = [0.94444+0.1134j, 0.94444-0.1134j]

K = control.acker(ss.A, ss.B, poles)

print('Ackermann matrix')
print(K)
```

As we obtain the gain matrix, we can form the regulator structure. Then, we can simulate the system to various initial conditions. Next, we provide such an example.

Example 10.2 *Obtaining output of the DC motor with regulator*

We can form the regulator structure for the DC motor based on the calculated gain matrix. To do so, we can benefit from the Python code given in Listing 10.2. This code provides output and state values of the DC motor to an initial condition with regulator applied to it.

Listing 10.2: Obtaining output and states of the DC motor in Python with regulator applied to it.

```
import mpcontrolPC
import mpcontrol
import control

num = [165702.47]
den = [1, 390.2, 15701]

tfs = control.tf(num, den)

Ts = 0.0005
tfz = control.c2d(tfs, Ts)

ss = control.tf2ss(tfz)

ss_cf,T_cf = control.canonical_form(ss,'reachable')

ss_mp = mpcontrolPC.ss_to_mpss(ss_cf)

F=ss_mp.A
G=ss_mp.B
C=ss_mp.C
D=ss_mp.D

poles = [0.94444+0.1134j, 0.94444-0.1134j]

Kx = control.acker(F, G, poles)
K = Kx.tolist()

x0 = [[10], [10]]

N = 1000

x1 = [0 for i in range(N)]
x2 = [0 for i in range(N)]
y = [0 for i in range(N)]
u = [0 for i in range(N)]

x = mpcontrol.create_empty_matrix(2, 1)
x_old = x0

x1[0] = x0[0][0]
x1[1] = x0[1][0]
x2[0] = x0[0][0]
x2[1] = x0[1][0]

for n in range(2, N):
    F_d = mpcontrol.mult_matrices(F, x_old)

    G_d2 = mpcontrol.mult_matrices(G, K)
    G_d1 = mpcontrol.mult_matrices(G_d2, x_old)
    G_d = mpcontrol.mult_matrix_scalar(G_d1, -1)

    x = mpcontrol.add_matrices(F_d, G_d)
```

```
K_d = mpcontrol.mult_matrices(K, x)
u[n] = -K_d[0][0]

C_d = mpcontrol.mult_matrices(C, x_old)
y[n] = C_d[0][0]
x_old = x
x1[n] = x[0][0]
x2[n] = x[1][0]
```

As we execute the Python code in Listing 10.2 on PC, we obtain the plots in Figure 10.7. As can be seen in this figure, we compute a state feedback matrix so that all initial conditions of the states, $x_1[n]$ and $x_2[n]$, are driven to zero as determined by the design specifications. Furthermore, the reference input is selected as zero and output, $y[n]$, goes to zero. It is possible for the reader to change the desired closed-loop poles for the pole placement method. Then, he or she can form another state feedback matrix to improve the transient response to make the system reach to zero faster.

10.3.2 Controller Design

In the controller design, we have the integrator added to the system. As explained in Section 10.2, we should modify the state-space representation accordingly.

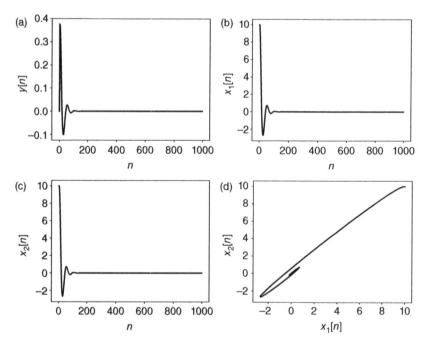

Figure 10.7 State-space and output plots of the DC motor with regulator applied to it. (a) Output. (b) First state. (c) Second state. (d) State space.

Hence, gain of the integrator is also calculated. We provide such an example next.

Example 10.3   *Finding the controller gain matrix for the DC motor*

We can design a controller for the DC motor using the Python control systems library. Hence, we will obtain the gain matrix in Eq. (10.11). To do so, we can modify the Python code in Listing 10.1. Assume that we want the closed-loop poles of the system to be located at $[0.9444+0.1134j, 0.9444-0.1134j]$, and -0.0003. We select the same desired closed-loop poles as in Example 10.1. Based on these values, we can use the Python code in Listing 10.3 on PC to calculate the state feedback (gain) matrix and integrator gain for the controller. As we execute the code, we obtain $\mathbf{K}_2 = \begin{bmatrix} -0.0694 & 0.0812 \end{bmatrix}$ and $k_1 = 0.0149$.

Listing 10.3: State feedback matrix and integrator gain for the controller.

```
import mpcontrolPC
import mpcontrol
import control

num = [165702.47]
den = [1, 390.2, 15701]

tfs = control.tf(num, den)

Ts = 0.0005
tfz = control.c2d(tfs, Ts)

ss = control.tf2ss(tfz)

ss_cf,T_cf = control.canonical_form(ss,'reachable')

ss_mp = mpcontrolPC.ss_to_mpss(ss_cf)

F=ss_mp.A
G=ss_mp.B
C=ss_mp.C
D=ss_mp.D

poles = [0.94444+0.1134j, 0.94444-0.1134j]
polesh = list.copy(poles)
polesh.append(-0.0003)

Fh = mpcontrol.create_empty_matrix(len(F)+1, len(F[0])+1)
for i in range(len(F)):
    for j in range(len(F[0])):
        Fh[i][j] = F[i][j]
for i in range(len(C[0])):
    Fh[len(F)][i] = -C[0][i]

Gh = mpcontrol.create_empty_matrix(len(G)+1, len(G[0]))
for i in range(len(G)):
    Gh[i][0] = G[i][0]
Gh[len(G)][0] = 0.0

Khx = control.place(Fh, Gh, polesh)
Kh = Khx.tolist()
K = list.copy(Kh)
```

```
k1 = -Kh[0][len(Kh[0])-1]
del K[0][len(K[0])-1]

print(Kh)
print(k1)
```

We can design a controller for the DC motor using the pole placement method. Since our DC motor is second degree, we can benefit from its standard representation given in Ogata (2009). This leads to projecting the desired time-domain requirements to corresponding closed-loop system poles. Let us consider such an example.

Example 10.4 *From transient response values to desired poles for the second-order system*

The continuous-time second-order system has a standard transfer function representation. Moreover, if the system is underdamped, its transient-response characteristics can be formulated as explained in detail in the book by Ogata (2009). Let us design a system with $T_{st} < 0.01$ seconds and $M_p < 10\%$. Based on these values, we will obtain $\zeta > 0.59$ and $\zeta\omega_n > 300$. If we select $\zeta = 0.60$, then we can set $\omega_n = 550$. These values are for the continuous-time system. Note that, selection of these parameters is important because they determine the system response. We can benefit from the methods introduced in Chapter 5 to obtain the corresponding digital system. Hence, we can obtain the desired poles to be used in the controller design in digital domain. We can use the Python code given in Listing 10.4 on PC for this purpose. As we execute the code, we obtain the desired poles as $z_1 = 0.8287 + 0.1858j$ and $z_1 = 0.8287 - 0.1858j$.

Listing 10.4: Calculating the desired poles in Python for the second order system.

```
import control

# natural frequency, damping ratio
wn = 550; zeta = 0.6; pole3=-0.00001

Ts=0.0005

num=[wn]
den=[1, 2*zeta*wn, wn**2]

gs = control.TransferFunction(num, den)

gz = control.sample_system(gs, Ts, method='bilinear')

poles=gz.pole()

#controller, modified
polesh = [poles[0],poles[1], pole3]

print(polesh)
```

We can benefit from the previous example to design a controller for the DC motor. Let us consider it next.

Example 10.5 *Obtaining output of the DC motor with controller*

We can form the controller structure for the DC motor based on the calculated state feedback matrix. To do so, we can benefit from the Python code given in Listing 10.5. This code provides output and state values of the DC motor to a unit-step input with controller applied to it. Here, we also set the third pole as pole3=-0.003.

Listing 10.5: Obtaining the output and state of the DC motor to the unit-step input with controller applied to it.

```
import mpcontrolPC
import mpcontrol
import control

num = [165702.47]
den = [1, 390.2, 15701]

tfs = control.tf(num, den)

Ts = 0.0005
tfz = control.c2d(tfs, Ts)

ss = control.tf2ss(tfz)

ss_cf, T_cf = control.canonical_form(ss, 'reachable')

ss_mp = mpcontrolPC.ss_to_mpss(ss_cf)

F = ss_mp.A
G = ss_mp.B
C = ss_mp.C
D = ss_mp.D

poles = [0.94444 + 0.1134j, 0.94444 - 0.1134j]
polesh = list.copy(poles)
polesh.append(-0.0003)

Fh = mpcontrol.create_empty_matrix(len(F) + 1, len(F[0]) + 1)
for i in range(len(F)):
    for j in range(len(F[0])):
        Fh[i][j] = F[i][j]
for i in range(len(C[0])):
    Fh[len(F)][i] = -C[0][i]

Gh = mpcontrol.create_empty_matrix(len(G) + 1, len(G[0]))
for i in range(len(G)):
    Gh[i][0] = G[i][0]
Gh[len(G)][0] = 0.0

Khx = control.place(Fh, Gh, polesh)
Kh = Khx.tolist()
K = list.copy(Kh)
k1 = -Kh[0][len(Kh[0]) - 1]
del K[0][len(K[0]) - 1]
```

```
N = 1000

r = mpcontrol.step_signal(N, 1)

x0 = [[0], [0]]

x1 = [0 for i in range(N)]
x2 = [0 for i in range(N)]
y = [0 for i in range(N)]
u = [0 for i in range(N)]
e = [0 for i in range(N)]
ei = [0 for i in range(N)]

x = mpcontrol.create_empty_matrix(2, 1)
x_old = x0

x1[0] = x0[0][0]
x1[1] = x0[1][0]
x2[0] = x0[0][0]
x2[1] = x0[1][0]

for n in range(2, N):
    e[n] = r[n] - y[n - 1]
    ei[n] = e[n] + ei[n - 1]

    K_d = mpcontrol.mult_matrices(K, x_old)
    u[n] = k1 * ei[n] - K_d[0][0]

    F_d = mpcontrol.mult_matrices(F, x_old)
    G_d = mpcontrol.mult_matrix_scalar(G, u[n])
    x = mpcontrol.add_matrices(F_d, G_d)

    C_d = mpcontrol.mult_matrices(C, x_old)
    y[n] = C_d[0][0]
    x_old = x
    x1[n] = x[0][0]
    x2[n] = x[1][0]
```

As we execute the Python code in Listing 10.5 on PC, we obtain the plots in Figure 10.8. The unit-step response of the DC motor is as in Figure 10.8(a). As can be seen in this figure, the system follows the step input well. It is again possible for the designer to change the desired closed-loop poles for the pole placement method to improve the unit-step response of the DC motor (such as faster response or less overshoot).

10.4   State Observer Design

Regulator and controller design by the pole placement method is based on state values representing the LTI system. Therefore, the system should be observable such that states can be measured and used in the feedback operation. There may be cases in which these states cannot be directly measured or reached. Or, cost of measuring these states may be high such that it becomes infeasible to directly measure them by sensors. For such cases, state observers can be used.

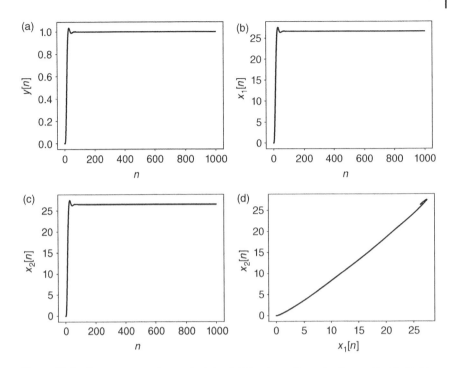

Figure 10.8 State-space and output plots of DC motor with controller applied to it. (a) Output. (b) First state. (c) Second state. (d) State space.

The observer estimates states $\mathbf{x}[n]$ of the LTI system from its input $u[n]$ and output $y[n]$ in the state-space representation. The basic idea behind the state observer is to place the model of the system in parallel with the actual system and drive both with the same input. Assuming that the estimated states follow the actual ones, we can use them in regulator and controller design afterward. We will next provide mathematical derivation of the state observer followed by its usage in code form.

10.4.1 Mathematical Derivation

Assume that we represent the estimated state value for $\mathbf{x}[n]$ as $\hat{\mathbf{x}}[n]$. There are always uncertainties in the system model in real world. Hence, $\hat{\mathbf{x}}[n]$ will diverge from $\mathbf{x}[n]$. The observer will have the same input as in the actual the system. The only way to compare the actual and observed states are via the measured and estimated outputs of the system as $y[n]$ and $\hat{y}[n] = \mathbf{C}\hat{\mathbf{x}}[n]$, respectively. We can use these values to correct the estimated state $\hat{\mathbf{x}}[n]$ and cause $\hat{\mathbf{x}}[n]$ to approach the actual state $\mathbf{x}[n]$. Then, the state-space equations for the estimated state and output values will be as, Ogata (1995),

$$\hat{x}[n+1] = F\hat{x}[n] + Gu[n] + L(y[n] - \hat{y}[n]) \tag{10.13}$$

$$\hat{y}[n] = C\hat{x}[n] \tag{10.14}$$

where **L** is the observer gain matrix that is determined as part of the observer design procedure. The state estimator error $e[n+1]$ is defined as

$$
\begin{aligned}
e[n+1] &= x[n+1] - \hat{x}[n+1] \\
&= (Fx[n] + Gu[n]) - (F\hat{x}[n] + Gu[n] + L(y[n] - \hat{y}[n])) \\
&= (Fx[n] + Gu[n]) - (F\hat{x}[n] + Gu[n] + L(Cx[n] - C\hat{x}[n])) \\
&= F(x[n] - \hat{x}[n]) - LC(x[n] - \hat{x}[n]) \\
&= (F - LC)(x[n] - \hat{x}[n]) \\
&= (F - LC)e[n]
\end{aligned}
\tag{10.15}
$$

As can be seen in Eq. 10.15, the state observer error equation has the same structure with the one we obtained in the regulator design. Hence, the error dynamics are obtained from $|zI - F + LC| = 0$. The main idea here is to make the state estimates, $\hat{x}[n]$, to converge to the actual state values, $x[n]$. To do so, we can follow the pole placement method to obtain **L**. As in the regulator design, we should define desired poles of the observer such that the observation operation is performed as expected. Ackermann's formula can be used in the observer design as well.

10.4.2 Ackermann's Formula for the Observer Gain

Assume that the solution of $|zI - F + LC| = 0$ is the set $\{\lambda_1, \dots, \lambda_m\}$. Hence, we can write $\alpha_e(z) = (z - \lambda_1) \cdots (z - \lambda_m) = 0$. It is possible to place the poles of the observer anywhere as long as the system is observable through the output $y[n]$ by selecting **L**. Then, Ackermann's formula for the observer gain matrix **L** becomes

$$
L = \alpha_e(F)
\begin{bmatrix}
C \\
CF \\
\cdots \\
CF^{m-1}
\end{bmatrix}
\begin{bmatrix}
0 \\
0 \\
\cdots \\
1
\end{bmatrix}
\tag{10.16}
$$

Using the calculated **L** value in Eq. (10.13), we can estimate the state values for the LTI system at hand. This leads to the regulator and controller design based on the observed state values. We will introduce these next in code form.

10.5 Regulator and Controller Design in Python using Observers

We will handle the regulator and controller design methods using observers in this section. To do so, we will first provide the Python code for state estimation. Then, we will use this setup in connection with the Python codes for the regulator and controller design methods introduced in Section 10.3.

10.5.1 Observer Design

As explained in Section 10.4, state observer can be constructed using Ackermann's formula. Therefore, we can use one of the functions `acker()` or `place()` within the Python control systems library. For more information on these functions, please see the library website. We will be using the function `acker()` to obtain the observer gain matrix **L**. However, the input matrix and pole values will be different for the function. For the observer case, we will have `acker(F_tran, C_tran, p)`. We will need the transposed form of matrices F and C for this purpose. The MicroPython control systems library has the function `transpose()` for this purpose. Therefore, we should have code lines `F_tran = mpcontrolPC.transpose(F)` and `C_tran = mpcontrolPC.transpose(C)` beforehand. Poles of the observer system can be different from the poles used in the regulator or controller design. The only rule here is that the observer should be fast in estimating the state values. We will provide such an example in this section. By the way, the poles used in the regulator or controller can also be used in the observer design as well. Next, we provide an example on calculating the state observer matrix via Ackermann's formula.

Example 10.6 *Finding the observer gain matrix for the DC motor*
We can design an observer for the DC motor using the Python control systems library. Hence, we will obtain the gain matrix **L**. Assume that, we want the poles of the observer for the DC motor to be located at [0.8, 0.79]. Based on these values, we can use the Python code in Listing 10.6 on PC to calculate the observer gain matrix. As we execute the code, we obtain the state feedback matrix as **L** = [6.0551 6.1308].

Listing 10.6: Finding the observer gain matrix bf L of the DC motor.

```
import mpcontrolPC
import control

num = [165702.47]
den = [1, 390.2, 15701]
```

```
tfs = control.tf(num, den)

Ts = 0.0005
tfz = control.c2d(tfs, Ts)

ss = control.tf2ss(tfz)

ss_cf,T_cf = control.canonical_form(ss,'reachable')

ss_mp = mpcontrolPC.ss_to_mpss(ss_cf)

F=ss_mp.A
G=ss_mp.B
C=ss_mp.C
D=ss_mp.D

p = [0.8, 0.79]

F_tran = mpcontrolPC.transpose(F)
C_tran = mpcontrolPC.transpose(C)

Lx = control.place(F_tran, C_tran, p)
L_tran = Lx.tolist()
L = mpcontrolPC.transpose(L_tran)

print('Observer gain')
print(L)
```

10.5.2 Observer-Based Regulator Design

The Python code introduced in Section 10.3 can be modified to include the observer in its operation. To do so, we should estimate the states first. Then, we should use these estimated values in the regulator structure. Hence, the observer-based regulator design will have two separate but connected steps as observer design and regulator formation based on it. We provide such an example next.

Example 10.7 *Observer-based regulator design for the DC motor in Python*
As in Example 10.2, we want to form a regulator for the DC motor. However, we will use the observed (estimated) state values in operation. To do so, we can benefit from the Python code in Listing 10.7. This code provides the simulated and estimated output and state values of the DC motor to an initial condition with an observer-based regulator applied to it.

Listing 10.7: Obtaining the output and states of the DC motor with observer and regulator applied to it.

```
import mpcontrolPC
import mpcontrol
import control

num = [165702.47]
```

```
den = [1, 390.2, 15701]

tfs = control.tf(num, den)

Ts = 0.0005
tfz = control.c2d(tfs, Ts)

ss = control.tf2ss(tfz)

ss_cf, T_cf = control.canonical_form(ss, 'reachable')

ss_mp = mpcontrolPC.ss_to_mpss(ss_cf)

F = ss_mp.A
G = ss_mp.B
C = ss_mp.C
D = ss_mp.D

poles = [0.94444 + 0.1134j, 0.94444 - 0.1134j]

Kx = control.place(F, G, poles)
K = Kx.tolist()

p = [0.8, 0.79]

F_tran = mpcontrolPC.transpose(F)
C_tran = mpcontrolPC.transpose(C)

Lx = control.place(F_tran, C_tran, p)
L_tran = Lx.tolist()
L = mpcontrolPC.transpose(L_tran)

N = 1000

x1 = [0 for i in range(N)]
x2 = [0 for i in range(N)]
y = [0 for i in range(N)]
u = [0 for i in range(N)]

x0 = [[10], [10]]

x = mpcontrol.create_empty_matrix(2, 1)
x_old = x0

x1[0] = x0[0][0]
x1[1] = x0[1][0]
x2[0] = x0[0][0]
x2[1] = x0[1][0]

x1_est = [0 for i in range(N)]
x2_est = [0 for i in range(N)]
y_est = [0 for i in range(N)]

x_est = mpcontrol.create_empty_matrix(2, 1)

x_est0 = [[0], [0]]
x_est_old = x_est0

x1_est[0] = x_est0[0][0]
x1_est[1] = x_est0[1][0]
x2_est[0] = x_est0[0][0]
x2_est[1] = x_est0[1][0]
```

```
for n in range(2, N):
    K_d = mpcontrol.mult_matrices(K, x_old)
    u[n] = -K_d[0][0]

    F_d = mpcontrol.mult_matrices(F, x_old)
    G_d = mpcontrol.mult_matrix_scalar(G, u[n])
    x = mpcontrol.add_matrices(F_d, G_d)

    C_d = mpcontrol.mult_matrices(C, x_old)
    y[n] = C_d[0][0]

    x_old = x
    x1[n] = x[0][0]
    x2[n] = x[1][0]

    # estimated states and output

    K_d = mpcontrol.mult_matrices(K, x_est_old)
    u[n] = -K_d[0][0]

    F_d = mpcontrol.mult_matrices(F, x_est_old)
    G_d = mpcontrol.mult_matrix_scalar(G, u[n])
    G_d2 = mpcontrol.add_matrices(F_d, G_d)
    L_d1 = y[n - 1] - y_est[n - 1]
    L_d = mpcontrol.mult_matrix_scalar(L, L_d1)
    x_est = mpcontrol.add_matrices(L_d, G_d2)

    C_d = mpcontrol.mult_matrices(C, x_est_old)
    y_est[n] = C_d[0][0]

    x_est_old = x_est
    x1_est[n] = x_est[0][0]
    x2_est[n] = x_est[1][0]
```

As we execute the Python code in Listing 10.7 on PC, we obtain the plots in Figure 10.9. As can be seen in this figure, the observer estimates converge to actual state variables quickly and track the states fairly well. We compute the state feedback matrix so that all estimated states are driven to zero. Furthermore, the reference input is zero. The output, $y[n]$, also goes to zero using the estimated states.

10.5.3 Observer-Based Controller Design

The Python code introduced in Section 10.3 can be modified to include the observer in its operation. To do so, we should estimate the states first. Then, we should use these estimated values in the controller structure. Hence, the observer-based controller design will have two separate but connected steps as observer design and controller formation based on it. We provide such an example next.

Example 10.8 *Observer-based controller design for the DC motor in Python*
As in Example 10.5, we want to form a controller for the DC motor. However, we will use the observed (estimated) state values in operation. To do so, we can benefit from the Python code given in Listing 10.8. This code provides the simulated

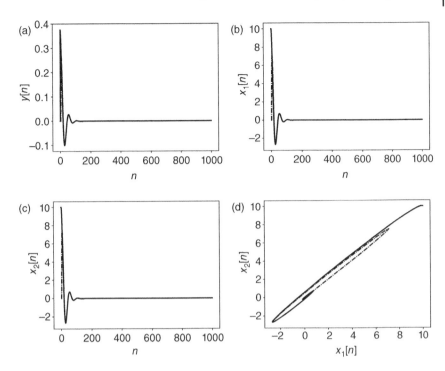

Figure 10.9 State-space and output plots of DC motor with observer and regulator applied to it. (a) Output and its estimate. (b) First state and its estimate. (c) Second state and its estimate. (d) State space.

and estimated output and state values of the DC motor to the step input with an observer-based controller applied to it.

Listing 10.8: Obtaining output and states of the DC motor in Python with observer and controller applied to it.

```
import mpcontrolPC
import mpcontrol
import control

num = [165702.47]
den = [1, 390.2, 15701]

tfs = control.tf(num, den)

Ts = 0.0005
tfz = control.c2d(tfs, Ts)

ss = control.tf2ss(tfz)

ss_cf, T_cf = control.canonical_form(ss, 'reachable')

ss_mp = mpcontrolPC.ss_to_mpss(ss_cf)
```

```python
F = ss_mp.A
G = ss_mp.B
C = ss_mp.C
D = ss_mp.D

poles = [0.94444 + 0.1134j, 0.94444 - 0.1134j]
polesh = list.copy(poles)
polesh.append(-0.0003)

Fh = mpcontrol.create_empty_matrix(len(F) + 1, len(F[0]) + 1)
for i in range(len(F)):
    for j in range(len(F[0])):
        Fh[i][j] = F[i][j]
for i in range(len(C[0])):
    Fh[len(F)][i] = -C[0][i]

Gh = mpcontrol.create_empty_matrix(len(G) + 1, len(G[0]))
for i in range(len(G)):
    Gh[i][0] = G[i][0]
Gh[len(G)][0] = 0.0

Khx = control.place(Fh, Gh, polesh)
Kh = Khx.tolist()
K = list.copy(Kh)
k1 = -Kh[0][len(Kh[0]) - 1]
del K[0][len(K[0]) - 1]

# observer part

p = [0.8, 0.79]

F_tran = mpcontrolPC.transpose(F)
C_tran = mpcontrolPC.transpose(C)

Lx = control.place(F_tran, C_tran, p)
L_tran = Lx.tolist()
L = mpcontrolPC.transpose(L_tran)

N = 1000

r = mpcontrol.step_signal(N, 1)
# r = mpcontrol.ramp_signal(N, .1)
# r = mpcontrol.sinusoidal_signal(N, .3, N/20, 0, 0, N, 1)


x1 = [0 for i in range(N)]
x2 = [0 for i in range(N)]
y = [0 for i in range(N)]
u = [0 for i in range(N)]
e = [0 for i in range(N)]
ei = [0 for i in range(N)]

x0 = [[10], [10]]

x = mpcontrol.create_empty_matrix(2, 1)
x_old = x0

x1[0] = x0[0][0]
x1[1] = x0[1][0]
x2[0] = x0[0][0]
x2[1] = x0[1][0]
```

```
x1_est = [0 for i in range(N)]
x2_est = [0 for i in range(N)]
y_est = [0 for i in range(N)]

x_est = mpcontrol.create_empty_matrix(2, 1)

x_est0 = [[0], [0]]
x_est_old = x_est0

x1_est[0] = x_est0[0][0]
x1_est[1] = x_est0[1][0]
x2_est[0] = x_est0[0][0]
x2_est[1] = x_est0[1][0]

for n in range(2, N):
    e[n] = r[n] - y[n - 1]
    ei[n] = e[n] + ei[n - 1]

    K_d = mpcontrol.mult_matrices(K, x_old)
    u[n] = k1 * ei[n] - K_d[0][0]

    F_d = mpcontrol.mult_matrices(F, x_old)
    G_d = mpcontrol.mult_matrix_scalar(G, u[n])
    x = mpcontrol.add_matrices(F_d, G_d)

    C_d = mpcontrol.mult_matrices(C, x_old)
    y[n] = C_d[0][0]

    x_old = x
    x1[n] = x[0][0]
    x2[n] = x[1][0]

    # estimated states and output

    K_d = mpcontrol.mult_matrices(K, x_est_old)
    u[n] = k1 * ei[n] - K_d[0][0]

    F_d = mpcontrol.mult_matrices(F, x_est_old)
    G_d = mpcontrol.mult_matrix_scalar(G, u[n])
    G_d2 = mpcontrol.add_matrices(F_d, G_d)
    L_d1 = y[n - 1] - y_est[n - 1]
    L_d = mpcontrol.mult_matrix_scalar(L, L_d1)
    x_est = mpcontrol.add_matrices(L_d, G_d2)

    C_d = mpcontrol.mult_matrices(C, x_est_old)
    y_est[n] = C_d[0][0]

    x_est_old = x_est
    x1_est[n] = x_est[0][0]
    x2_est[n] = x_est[1][0]
```

As we execute the Python code in Listing 10.8 on PC, we obtain the plots in Figure 10.10. More specifically, unit-step response of the DC motor is as in Figure 10.10(a). As can be seen in this figure, the controller formed by the estimated states lead the DC motor to track the input signal fairly well. Thus, the designed state observer satisfies the desired specification of the system response defined by the closed-loop poles.

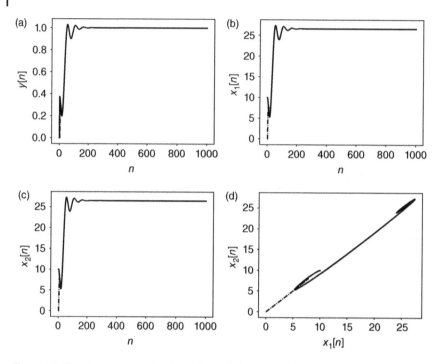

Figure 10.10 State-space and output plots of DC motor with observer and controller applied to it. (a) Output and its estimate. (b) First state and its estimate. (c) Second state and its estimate. (d) State space.

10.6 Application: State-space based Control of the DC Motor

In this application, we will introduce methods to implement an observer-based controller on the STM32 microcontroller. The hardware setup in Section 7.4 will be used in this application. The only difference here is that we will use a different controller structure.

10.6.1 Hardware Setup

As mentioned before, we will use the hardware setup in Section 7.4 in this application. We will benefit from the quadrature encoder output while forming the feedback control signal.

10.6.2 Procedure

Here, we control the DC motor using observer-based state-space controller. We will use the **K** and **L** matrices obtained in Listing 10.8 for this purpose. We apply a constant amplitude input with 60 rpm to the DC motor. Then, we obtain the output and estimate of the first state variable. We send these to PC in real-time for one second. Afterward, we use the discrete-time state-space equations of the DC motor obtained in Chapter 9. We simulate the system using these equations with the same input signal. Finally, we compare the actual system and simulation results.

10.6.3 C Code for the System

For the application, we form a new project in Mbed Studio and use the C code in Listing 10.9. The software setup will be the same as in Section 3.4.3. As we execute this code, observer-based state-space controller is constructed using the function `controller_ss_init(float *F, float *G, float *C, float *D, float *K, float *L, float *x_est0, float k1, int no_states)`. Here, F, G, C, and D are state-space matrices of the system. K, L, and k1 are the gain and observer matrices. x_est0 is the initial condition of the estimated states. no_states is the state number. The function `send_data_RT(int select)` is used to send output and state variable 1 signals to PC in real-time. Here, select is chosen as 3 for this purpose.

Listing 10.9: The C code to be used in implementing observer-based state-space controller for the DC motor.

```
#include "mbed.h"
#include "system_general.h"
#include "system_input.h"
#include "system_output.h"
#include "system_controller.h"

Serial pc(USBTX, USBRX, 115200);
RawSerial device(PD_5, PD_6, 921600);
Ticker sampling_timer;
InterruptIn button(BUTTON1);
Timer debounce;
DigitalOut led(LED1);
PwmOut pwm1(PE_9);
PwmOut pwm2(PE_11);
DigitalOut pwm_enable(PF_15);
InterruptIn signal1(PA_0);
InterruptIn signal2(PB_9);
Timer encoder_timer;
AnalogIn ADCin(PA_3);
AnalogOut DACout(PA_4);

#define no_states 2
```

```
float F[no_states][no_states] = {
    {1.8191874693263341, -0.822752378704249},
    {1.0, 0.0}
};

float G[no_states][1] = {
    {1.0},
    {0.0}
};

float C[1][no_states] = {
    {0.01942273369018399, 0.018199983923132712}
};

float D[1][1] = {
    {0.0}
};

float K[1][no_states] = {
{-0.06939253067366513, 0.08121774596459652}
};

float L[no_states][1] = {
{6.055136683414969},
{6.130783551125799}
};

float k1 = 0.014914735267154897;

float xest0[no_states][1] = {
    {0.0},
    {0.0}
};

int main()
{
    user_button_init();
    system_input_init(1, 12/3.3);
    system_output_init(1, 3591.84);
    send_data_RT(3);
    sampling_timer_init();
    controller_ss_init(&F[0][0], &G[0][0], &C[0][0], &D[0][0],
        &K[0][0], &L[0][0], &xest0[0][0], k1, no_states);
    menu_init();
    while (true) {
        system_loop();
    }
}
```

In the system_controller.cpp file, the function obtain_controller_ output(float input, float output, int cnt_signal) is used to implement observer-based state-space controller. A flag inside this function separates transfer function based controller and observer-based state-space controller implementation. This function takes the desired reference as input and previous output data as output. It creates output of the observer-based state-space controller. This output is used as the system input. In Listing 10.9, we send output and state variable 1 signals to PC for further processing. On the

PC side, we benefit from the Python code in Listing 10.10 for this purpose. Here, we use the function `get_data_from_MC(NoElements, com_port, selection)` to receive both the state variable and output signals. As we receive these values, we can save them to separate files.

Listing 10.10: The Python code to receive and store state and output values.

```
import mpcontrol
import mpcontrolPC
import matplotlib.pyplot as plt

N = 2000
output_signal, state1_est = mpcontrolPC.get_data_from_MC
(N, 'com10', 1)

mpcontrolPC.save_to_file(output_signal, file_name='output.dat')
mpcontrolPC.save_to_file(state1_est, file_name='state1_est.dat')
```

We should use the same controller in Listing 10.8 to simulate the system. To do so, we should modify the parts of the code. First, the N value should be set to 2000 since we will compare this result with the actual data received for one second. Next, the step input signal should have amplitude 60. Again, this is necessary to compare the simulation and actual data. As we perform these two modifications, we can compare the actual and simulated results on PC.

10.6.4 Python Code for the System

We can repeat this application in MicroPython on the STM32 microcontroller. The reader can use the Python code in Listing 10.11 for this application. Obtaining data, plotting result, and saving data to files are same as explained in Section 10.6.3.

Listing 10.11: The Python code to be used in implementing observer-based state-space controller for the DC motor.

```
import system_general
import system_input
import system_output
import system_controller

def main():
        F = [[1.8191874693263341, -0.822752378704249], [1.0, 0.0]]
        G = [[1.0], [0.0]]
        C = [[0.01942273369018399, 0.018199983923132712]]
        D = [[0.0]]
        K = [[-0.06939253067366513, 0.08121774596459652]]
        L = [[6.055136683414969], [6.130783551125799]]
        x_est0 = [[0.0], [0.0]]
        k1 = 0.014914735267154897
        Ts = 0.0005

        system_general.button_init()
        system_general.PWM_frequency_select(0)
        system_general.sampling_timer_init(0)
        system_general.input_select(input_type='Constant Amplitude',
```

```
                amplitude=60)
                system_general.motor_direction_select(1)
                system_input.system_input_init(2, 3.63636363636)
                system_output.system_output_init(1, 3591.84)
                system_controller.controller_ss_init(F, G, C, D, Ts, K, L,
                    k1, x_est0)
                system_general.send_data_RT(2)
                while True:
                        system_general.system_loop()
main()
```

10.6.5 Observing Outputs

We provide the actual and simulated output signals in Figure 10.11. We also provide the actual and simulated first state value wrt time in the same figure. When
we compare these results, we can see that steady-state values for both methods are
the same except for fluctuations in actual values. Noise in the DC motor quadrature encoder reading causes these fluctuations. On the other hand, the actual and
simulated output values are different. There are two reasons for this difference.
First, we used the transfer function obtained in Chapter 6 for simulation. This

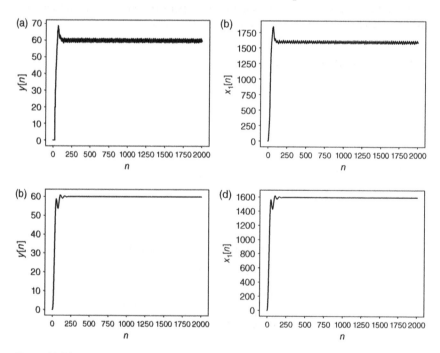

Figure 10.11 Actual signals obtained by the C code and simulated signals obtained by
Python code. (a) Actual output. (b) Actual first state. (c) Simulated output. (d) Simulated
first state.

transfer function is not identified perfectly. Second, output is limited between 0 and 12 V in the system implemented on the STM32 microcontroller. However, there is no such restriction when we perform the simulation. Therefore, there is a difference between the actual and simulated values.

When we execute the Python code and obtain the plots as in Figure 10.11, we can see that there are some minor differences compared to the C implementation. The reason for this is as follows. We use pin interrupts and timers to read speed information from the quadrature encoder in C implementation. On the other hand, we use the timer capture mode to obtain the same information in MicroPython implementation. This causes the noise in the quadrature encoder readings, hence the difference in output signals.

10.7 Summary

We considered state-space based controller design methods in this chapter. We approached the problem in two parts as regulator and controller design. To do so, we benefit from the pole placement method. We also introduced the state observer concept for cases in which the actual value of a state cannot be measured directly. We introduced methods to design regulators and observers based on actual state measurements and observer-based state estimation. We provided Python- and C-based functions to implement the designed controllers on the STM32 microcontroller. To do so, we had an end of chapter application summarizing all implementation steps.

Problems

10.1 Consider the DC motor with transfer function

$$G(s) = \frac{165702.47}{s^2 + 390.2s + 15701} \tag{10.17}$$

a. convert the system to digital form using bilinear transformation with $T_s = 0.0005$ seconds.
b. calculate the regulator gain matrix for the digital system with desired closed loop poles at [0.84+0.28j, 0.84-0.28j].

10.2 Consider the digital system in Problem 10.1, design a regulator for it.
a. Find output of the system with the regulator for initial conditions of states at $x_1[0] = 1$ and $x_2[0] = 1$.

b. Find output of the system with the regulator for initial conditions of states at $x_1[0] = 10$ and $x_2[0] = 10$.

10.3 Find the state feedback controller and integrator constant of the digital system in Problem 10.1 with the desired closed loop poles at [0.84+0.28j, 0.84-0.28j] and -0.008.

10.4 Consider the digital system in Problem 10.1,
a. calculate the desired closed loop poles of a second-order system with transient response characteristics $T_s < 0.005$ and $M_p < 20\%$.
b. obtain output and states of the system for the unit-step input with the desired closed-loop poles found in part (a).
c. obtain output and states of the system for the ramp input with the desired closed loop poles found in part (a).
d. obtain output and states of the system for the sinusoidal input with the desired closed loop poles found in part (a).

10.5 Consider the RC filter with transfer function

$$G(s) = \frac{1000}{s + 1000} \tag{10.18}$$

a. convert the system to digital form using bilinear transformation with $T_s = 0.0005$ seconds.
b. calculate the regulator gain matrix for this system with the desired closed loop pole at 0.85.

10.6 Consider the digital system in Problem 10.5, design a regulator for it.
a. Find output of the system with the regulator for initial conditions of the states at $x_1[0] = 1$ and $x_2[0] = 1$.
b. Find output of the system with the regulator for initial conditions of the states at $x_1[0] = 10$ and $x_2[0] = 10$.

10.7 Find the state feedback controller and integrator constant of the digital system in Problem 10.5 with the desired closed-loop poles at 0.85, -0.008.

10.8 Find the observer gain matrix of the digital system in Problem 10.1 when poles of the observer are located at [0.6, 0.59].

10.9 Design a regulator for the digital system in Problem 10.1 using the calculated state feedback matrix in Problem 10.2 and the observer gain matrix obtained in Problem 10.8. Find output and states of the system with this state feedback matrix.

10.10 Consider the digital system in Problem 10.1,
  a. obtain output and states of the system for the unit-step input with the selected closed-loop poles used in Problem 10.3 and the observer gain matrix obtained in Problem 10.8.
  b. obtain output and states of the system for the sinusoidal input with the selected closed-loop poles used in Problem 10.3 and the observer gain matrix obtained in Problem 10.8.

10.11 Find the observer gain matrix for the digital system in Problem 10.5 when pole of the observer is located at 0.6.

10.12 Design a regulator for the digital system in Problem 10.5 using the calculated state feedback matrix in Problem 10.7 and the observer gain matrix obtained in Problem 10.11. Find output and states of the system with this state feedback matrix.

11

Adaptive Control

Adaptive control provides another perspective to design a controller for the system at hand. Therefore, it deserves special consideration. In this chapter, we will explain adaptive control concepts starting from general layout. Then, we will continue with parameter estimation. Afterward, we will introduce indirect self-tuning regulator (ISTR) and model-reference adaptive control (MRAC) as two different approaches in adaptive control. Finally, we will provide an end of chapter application as the real-time parameter estimation of a given system, the DC motor for our case.

11.1 What is Adaptive Control?

A well-designed closed-loop controller can work efficiently in most applications. However, there may be changes in the system or environment during operation. As the first case, system characteristics may change over time. So, it may not be possible to control it as desired. Therefore, controller parameters should be adjusted to compensate variations of the system. The first step here should be finding the new parameter values. To do so, parameter estimation can be applied on the fly. This is system identification, as explored in Section 11.2, now in real-time. Then, the controller can be modified for the changed parameters. As the second case, the environment may change. There, we may need to update the controller to compensate these changes. For both cases, the controller should adopt itself to changing conditions. Hence, we will have the adaptive controller.

We provide general layout of the adaptive control scheme in Figure 11.1. As can be seen in this figure, there are two loops in operation. One loop consists of feedback with the system and controller as in the closed-loop structure considered thus far. The second loop is the adjustment mechanism used to update controller parameters.

Embedded Digital Control with Microcontrollers: Implementation with C and Python,
First Edition. Cem Ünsalan, Duygun E. Barkana, and H. Deniz Gürhan.
© 2021 The Institute of Electrical and Electronics Engineers, Inc. Published 2021 by John Wiley & Sons, Inc.
Companion website: www.wiley.com/go/Unsalan/Embedded_Digital_Control_with_Microcontrollers

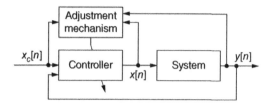

Figure 11.1 General layout of the adaptive control scheme.

There are various adaptive control methods such as gain scheduling, MRAC, ISTR, and dual control (Åström and Wittenmark 2013). These methods can be divided into two main categories as direct and indirect referring to the way controller parameters are updated. System parameters are estimated first in the indirect method. Then, controller parameters are obtained from the solution of a design problem using the estimated parameters. ISTR is one such method to be explored in Section 11.3. Controller parameters are updated directly in the direct method. MRAC is such a method to be explored in Section 11.4.

11.2 Parameter Estimation

An adaptive controller can be designed when parameters of the system to be controlled are known. If these parameters have changed or they are unknown, we can apply parameter estimation as the first step. Then, the controller can be designed using the estimated parameters.

There are several parameter estimation techniques in literature (Åström and Wittenmark 2013). In this book, we pick the recursive least squares (RLS) method. This method can be explained as follows. Let the LTI system with input $x[n]$ and output $y[n]$ be represented as

$$G(z) = \frac{z^{-d}B(z^{-1})}{A(z^{-1})} \tag{11.1}$$

where d is the delay parameter and

$$A(z^{-1}) = 1 + a_1 z^{-1} + a_2 z^{-2} \cdots + a_{n_a} z^{-n_a} \tag{11.2}$$

$$B(z^{-1}) = b_0 + b_1 z^{-1} + b_2 z^{-2} \cdots + b_{n_b} z^{-n_b} \tag{11.3}$$

We can represent the LTI system by its constant-coefficient difference equation as

$$
\begin{aligned}
y[n] = &- a_1 y[n-1] - a_2 y[n-2] \cdots - a_{n_a} y[n-n_a] \\
&+ b_0 x[n-d] + b_1 x[n-d-1] \cdots + b_{n_b} x[n-d-n_b]
\end{aligned} \tag{11.4}
$$

We can represent Eq. (11.4) as

$$y[n] = \boldsymbol{\varphi}^T[n]\hat{\theta}[n] \tag{11.5}$$

where

$$\boldsymbol{\varphi}^T[n] = [-y[n-1] \quad \cdots -y[n-n_a] \quad x[n-d] \quad \cdots \quad x[n-d-n_b]] \tag{11.6}$$

$$\hat{\theta}^T[n] = [a_1 \quad \cdots \quad a_{n_a} \quad b_0 \quad \cdots \quad b_{n_b}] \tag{11.7}$$

We can estimate $\hat{\theta}^T[n]$ iteratively using the RLS method as

$$\hat{\theta}[n+1] = \hat{\theta}[n] + K[n]\epsilon[n] \tag{11.8}$$

where

$$K[n] = P[n-1]\boldsymbol{\varphi}[n]\left(I + \boldsymbol{\varphi}^T[n]P[n-1]\boldsymbol{\varphi}[n]\right)^{-1}$$

$$\epsilon[n] = y[n] - \boldsymbol{\varphi}^T[n]\hat{\theta}[n-1]$$

$$P[n] = \frac{\left((I - K[n]\boldsymbol{\varphi}^T[n])P[n-1]\right) + \left((I - K[n]\boldsymbol{\varphi}^T[n])P[n-1]\right)^T}{2}$$

There are two functions in the MicroPython control systems library for parameter estimation via RLS. The first function `RLS_Sim(x, y, Nnum, Nden, Ts, gp)` provides the estimated parameters as well as their evolution in time. We should provide a known input signal and output obtained from it in array form as `x` and `y`. We should also feed the order of numerator and denominator polynomials to the function as `Nnum` and `Nden`, respectively. Finally, we should enter the sampling period `Ts` and gain value `gp` to the function. This gain value is used as a multiplier for the P matrix just for the first step in iteration. There is also the real-time version of the parameter estimation method by RLS in C language. We will introduce it in Section 11.5. While forming these functions, we are inspired by the MATLAB code by Attya at https://www.mathworks.com/matlabcentral/fileexchange/58121-recursive-least-square. Besides the RLS method, Attya also provides other adaptive control algorithms originated from the book by Åström and Wittenmark (2013). Therefore, we strongly suggest the reader to visit the mentioned web site.

The function `RLS_Sim` returns a class keeping all necessary information on system parameters as the estimation runs. Let us show how the function works on the DC motor with the next example.

Example 11.1 *Parameter estimation of the DC motor using the RLS method*
We extracted transfer function of the DC motor in Chapter 6 as in Eq. (6.18). We can use the Python code in Listing 11.1, to obtain the RLS estimate of the DC

motor parameters via simulation. To do so, we first construct the transfer function and feed a random signal to it. We obtain the output signal from the system. Then, we feed the input signal, corresponding output signal and necessary system parameters, such as the number of numerator and denominator coefficients, to the function RLS_Sim.

Listing 11.1: Estimating DC motor parameters using the RLS method.

```
import mpcontrolPC
import mpcontrol

Ts = 1
num = [0.00943, 0.01886, 0.00943]
den = [1, -1.8188, 0.8224]
g = mpcontrolPC.tf(num, den, Ts)

print(g)

N = 300
x = mpcontrol.random_signal(N, 1,200,0)
y = mpcontrol.lsim(g, x)

Ts = 0.0005

rls_simulation = mpcontrolPC.RLS_Sim(x, y, 2, 2, Ts, 1000000)

ge = rls_simulation.Sim()

print(ge)

#Plot section
import matplotlib.pyplot as plt

n=list(range(0, N))

plt.figure()
plt.plot(n, rls_simulation.num[0],'k')
plt.ylabel('b0')
plt.xlabel('n')

plt.figure()
plt.plot(n, rls_simulation.den[1],'k')
plt.ylabel('a1')
plt.xlabel('n')

plt.show()
```

As we execute Listing 11.1, we obtain rls_simulation which holds all the necessary information about the estimation process. Here, we first extract transfer function of the estimated system via rls_simulation.Sim() as

$$G(z) = \frac{0.0094z^2 + 0.0189z + 0.0094}{z^2 - 1.8187z + 0.8223} \tag{11.9}$$

As we compare Eqs. (6.18) and (11.9), we can see that the RLS method estimated the actual numerator and denominator coefficients fairly well. We can observe the iterative convergence of these parameters to their final value.

To do so, we should use the command `rls_simulation.num[·]` and `rls_simulation.den[·]`. We picked the coefficients b_0 and a_1 in Listing 11.1 for demonstration purposes. We plot the evolution of these two parameters in time in Figure 11.2. As can be seen in this figure, the two parameters converge to their final and true value fairly fast.

11.3 Indirect Self-Tuning Regulator

ISTR is the first adaptive controller design method to be considered in this chapter. In ISTR, system parameters are estimated first. Then, the controller is designed based on them. Thus, the controller parameters are updated indirectly via the estimated system parameters.

General block diagram for the ISTR is as in Figure 11.3. As can be seen in this figure, there are two modules as the feedforward controller with the transfer function $\frac{T(z^{-1})}{R(z^{-1})}$ and feedback controller with the transfer function $\frac{S(z^{-1})}{R(z^{-1})}$. In Figure 11.3, $G(z)$ is the transfer function of the system given in Eq. (11.1).

We can represent the numerator and denominator polynomials for $\frac{T(z^{-1})}{R(z^{-1})}$ and $\frac{S(z^{-1})}{R(z^{-1})}$ as

$$S(z^{-1}) = b_0 + b_1 z^{-1} + b_2 z^{-2} + \cdots + b_{n_b} z^{-n_b} \tag{11.10}$$

$$R(z^{-1}) = 1 + a_1 z^{-1} + a_2 z^{-2} + \cdots + a_{n_a} z^{-n_a} \tag{11.11}$$

$$T(z^{-1}) = c_0 + c_1 z^{-1} + c_2 z^{-2} + \cdots + c_{n_c} z^{-n_c} \tag{11.12}$$

The control law for the ISTR is $R(z^{-1})X(z) = T(z^{-1})X_c(z) - S(z^{-1})Y(z)$. Here, $X_c(z)$ is the reference input, $X(z)$ is the system input (control signal), and $Y(z)$ is output. During the controller design, we either know the system parameters or estimate them as in Section 11.2.

11.3.1 Feedback ISTR Design

We can design an ISTR by assuming that only the feedback controller, $\frac{S(z^{-1})}{R(z^{-1})}$, is available. Hence, transfer function of the feedforward controller in Figure 11.3 is set to unity for this setup. Then, the corresponding closed-loop transfer function becomes

$$\frac{Y(z)}{X_c(z)} = \frac{S(z^{-1})z^{-d}B(z^{-1})}{A(z^{-1})R(z^{-1}) + z^{-d}B(z^{-1})S(z^{-1})} \tag{11.13}$$

We can obtain the denominator polynomial of the transfer function in Eq. (11.13) as $A(z^{-1})R(z^{-1}) + z^{-d}B(z^{-1})S(z^{-1})$. As in the pole placement method

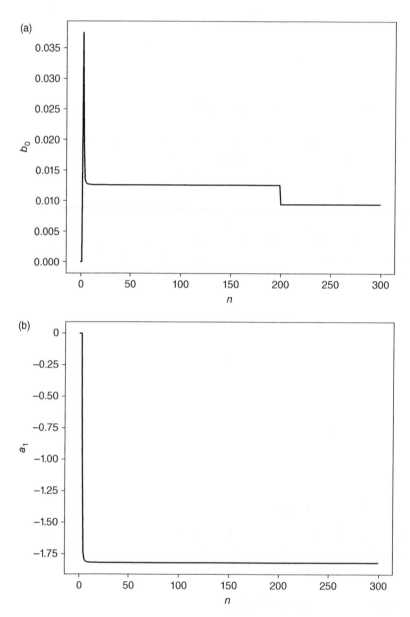

Figure 11.2 Estimation of the two system parameters in time. (a) b_0 wrt n. (b) a_1 wrt n.

Figure 11.3 General block diagram of ISTR.

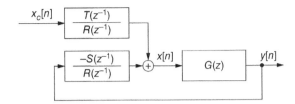

introduced in Section 10.2, we can equate the denominator polynomial to $\alpha(z^{-1}) = 1 + \alpha_1 z^{-1} + \alpha_2 z^{-2} \cdots + \alpha_{n_a} z^{-n_a}$ formed by the desired system and observer poles. Based on the equality $A(z^{-1})R(z^{-1}) + z^{-d}B(z^{-1})S(z^{-1}) = \alpha(z^{-1})$, we can find coefficients of the polynomials $S(z^{-1})$ and $R(z^{-1})$ from the coefficients of $\alpha(z^{-1})$.

In order to find the coefficients of the polynomials $S(z^{-1})$ and $R(z^{-1})$, we should use the Diophantine equation (Åström and Wittenmark 2013). To do so, we formed the function `Diophantine(g, d, Am_poles, AO_poles)` under the MicroPython control systems library. Here, g stands for the transfer function of the system to be controlled. d is the system delay parameter. Am_poles and AO_poles represent the desired system and observer poles, respectively. While forming this function, we are inspired by the MATLAB code by Attya. Let us demonstrate the usage of the feedback ISTR on the DC motor by an example next.

Example 11.2 *ISTR for the DC motor with feedback controller*

We can use the Python code in Listing 11.2 on PC to apply the ISTR with feedback controller to the DC motor. To do so, we first obtain the zero-order hold equivalent of the DC motor from its continuous-time transfer function. Then, we select the desired closed poles as $0.8 \pm j0.2$ and the desired observer pole as 0.5. Then, we obtain the controller coefficients using the Diophantine function.

Listing 11.2: ISTR for the DC motor with feedback controller.

```
import mpcontrolPC
import control

num=[165702.47]
den=[1, 390.2, 15701]

gs = control.TransferFunction(num, den)

#Transfer function of the system
Ts=0.0005
gz = control.sample_system(gs, Ts, method='zoh')
g = mpcontrolPC.tf_to_mptf(gz)

Am_poles = [0.8+0.2j, 0.8-0.2j]
AO_poles = [0.5]
d = 1
```

```
R, S = mpcontrolPC.Diophantine(g, d, Am_poles, A0_poles)
```

```
#Transfer function of the controller (S(z)/R(z))
h = mpcontrolPC.tf(S,R,Ts)
```

print(h)

```
# Transfer function of the closed-loop controlled system
gcl = mpcontrolPC.feedback(g,h)
```

As we execute the Python code in Listing 11.2, we obtain the ISTR with feedback controller as

$$\frac{S(z^{-1})}{R(z^{-1})} = \frac{3.7213z - 2.7194}{z - 0.3531} \tag{11.14}$$

We plot unit step response of the overall system as in Figure 11.4. We can add the code line `gcl.stepinfo()` to the Python code in Listing 11.2 to obtain time domain transient response measures of the overall system. As we execute this code, we obtain $T_{rt} = 0.0035$ seconds, $M_p = 7.5809\%$, $T_{pt} = 0.007$ seconds, and $T_{st} = 0.01$ second. These values are acceptable. We can also obtain the steady state error of the system as $e_{ss} = 0.3895$. Unfortunately, this value is not acceptable. Therefore, we will add the feedforward controller to ISTR in the next section.

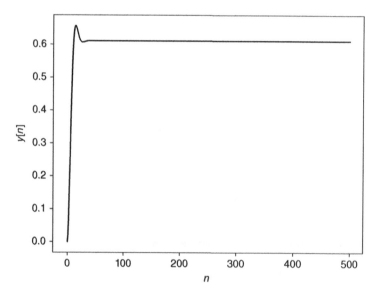

Figure 11.4 Unit step response of the DC motor with feedback controller in ISTR.

11.3.2   Feedback and Feedforward ISTR Design

We can add the feedforward controller besides the feedback one to ISTR. Then, closed-loop transfer function of the overall system becomes

$$\frac{Y(z)}{X_c(z)} = \frac{T(z^{-1})z^{-d}B(z^{-1})}{A(z^{-1})R(z^{-1}) + z^{-d}B(z^{-1})S(z^{-1})} \tag{11.15}$$

As can be seen in Eq. (11.15), the denominator polynomial for this case is the same as in Eq. (11.13). Hence, we obtain $S(z^{-1})$ and $R(z^{-1})$ using the Diophantine equation. Then, coefficients of $T(z^{-1})$ are obtained to satisfy the closed-loop response requirements (Åström and Wittenmark 2013).

Example 11.3   *ISTR of the DC motor with feedback and feedforward controllers.*
We can extend Example 11.2 by adding the feedforward controller to the ISTR design. We can use the Python code in Listing 11.3 on PC for this purpose. As in the previous example, we select the desired closed poles as $0.8 \pm j0.2$ and the desired observer pole as 0.5. Then, we obtain the controller coefficients using the Diophantine equation and closed-loop requirements. In this example, we benefit from Attya's code to obtain the $T(z^{-1})$ coefficients.

Listing 11.3: ISTR for the DC motor with feedback and feedforward controllers.

```
import mpcontrolPC
import control

num=[165702.47]
den=[1, 390.2, 15701]

gs = control.TransferFunction(num, den)

#Transfer function of the system
Ts=0.0005
gz = control.sample_system(gs, Ts, method='zoh')
g = mpcontrolPC.tf_to_mptf(gz)

Am_poles = [0.8+0.2j, 0.8-0.2j]
A0_poles = [0.5]
d = 1

R, S = mpcontrolPC.Diophantine(g, d, Am_poles, A0_poles)

#Transfer function of the controller (S(z)/R(z))
h = mpcontrolPC.tf(S,R,Ts)
#Transfer function of the controller (T(z)/R(z))
num3 = [-34.0225, 35.0857]
den3 = R

c3 = mpcontrolPC.tf(num3,den3,Ts)

# Transfer function of the closed-loop controlled system
gcl = mpcontrolPC.feedback(g,h)
gcl = mpcontrolPC.series(gcl,c3)
```

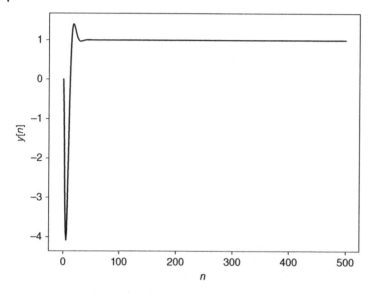

Figure 11.5 Unit step response of the DC motor with feedback and feedforward controllers in the ISTR.

As we execute the Python code in Listing 11.3, we obtain the ISTR feedback and feedforward controllers as

$$\frac{S(z^{-1})}{R(z^{-1})} = \frac{3.7213z - 2.7194}{z - 0.35309} \tag{11.16}$$

$$\frac{T(z^{-1})}{R(z^{-1})} = \frac{-34.0225z + 35.0857}{z - 0.3509} \tag{11.17}$$

We plot unit step response of the overall system as in Figure 11.5. We can add the code line `gcl.stepinfo()` to the Python code in Listing 11.3 to obtain transient response measures of the overall system. As we execute this code, we obtain $T_{rt} = 0.001$ second, $M_p = 40.2973\%$, $T_{pt} = 0.009$ seconds, and $T_{st} = 0.0175$ seconds. We can also obtain steady state error of the system as $e_{ss} = 4.6385e{-}06$. Compared to the feedback controller case, we can observe that performance of the system drastically improved. Only M_p, T_{pt}, and T_{st} values slightly increased.

11.4 Model-Reference Adaptive Control

MRAC is the second design method to be considered in this chapter. The main idea in MRAC is to choose a reference model to represent the desired system behavior. Then, the controller is set such that the overall system acts as the reference model. MRAC directly adjusts controller parameters for this purpose.

Figure 11.6 MRAC setup for gain adjustment.

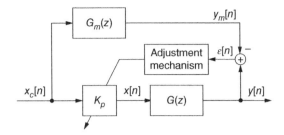

There are various MRAC setups for different systems (Åström and Wittenmark 2013). We pick the simplest setup as feedforward gain adjustment in this chapter. We assume the reference model and actual systems are the same. Only their gain parameters are different. MRAC is used to update the proportional controller gain. Hence, the system serially connected to the controller has the same characteristics with the reference model. We provide the layout for this operation in Figure 11.6.

In Figure 11.6, we assume $G_m(z) = K_m G(z)$ which is our reference model. Through the adjustment mechanism, we obtain K_p such that it converges to K_m. Hence, the system with the proportional controller will be the same as the reference model $G_m(z)$.

A suitable candidate for parameter adjustment is the MIT rule. We introduced this method for digital signal processing on Arm Cortex-M microcontrollers in our book (Ünsalan et al. 2018). With a minor modification for the setup in Figure 11.6, we can reach the iterative parameter adjustment method as

$$K_p[n+1] = K_p[n] - \gamma T_s \varepsilon[n] y_m[n] \tag{11.18}$$

where γ is the gain parameter for iteration, T_s is the sampling period. $\varepsilon[n]$ is the difference between the actual output and reference model output. $y_m[n]$ is the reference model output.

MicroPython control systems library has the function MRAC_Sim(x, ym, g, Ts, gamma) which provides the gain parameter K_p in Figure 11.6 and its evolution in time. We should provide a known input signal and output obtained from the reference model in array form as x and ym to the function. We should also feed the actual system model as g. Finally, we should enter the sampling period Ts and gain value gamma to the function. We explain working principles of the function MRAC_Sim on the DC motor by an example next.

Example 11.4 *MRAC for the DC motor*

We can use the Python code in Listing 11.4, to obtain the K_p value for the DC motor via MRAC setup in Figure 11.6. Here, we set $G_m(z) = 0.5G(z)$ and ask the adjustment mechanism in MRAC to adjust K_p to reach the final value 0.5.

Listing 11.4: MRAC for adjusting the K_p gain in DC motor.

```
import mpcontrolPC
import mpcontrol

Ts = 0.0005
num = [0.0094275, 0.018855, 0.0094275]
den = [1, -1.81883, 0.822399]
g = mpcontrol.tf(num, den, Ts)

num2 = [0.5]
den2 = [1]
g_gain = mpcontrol.tf(num2, den2, Ts)
g_model = mpcontrolPC.series(g_gain, g)

N = 2000
x = mpcontrol.step_signal(N, 1)
y_model = mpcontrol.lsim(g_model, x)

MRAC_simulation = mpcontrolPC.MRAC_Sim(x, y_model, g, Ts, 1)
MRAC_simulation.Sim()

#Plot section
import matplotlib.pyplot as plt

n=list(range(0, N))

plt.figure()
plt.plot(n, MRAC_simulation.theta,'k')
plt.ylabel('Kp')
plt.xlabel('n')
plt.show()
```

As we execute Listing 11.4, we obtain `MRAC_simulation.Sim()` which holds all the necessary information about the parameter adjustment process. We plot the evolution of K_p in time in Figure 11.7. As can be seen in this figure, the gain K_p converges to its true value (0.5 here) fairly fast.

11.5 Application: Real-Time Parameter Estimation of the DC Motor

The aim of this application is to use the RLS method in estimating parameters of the DC motor in real-time. Hence, we can construct transfer function of the system which can be used in adaptive controller implementations.

11.5.1 Hardware Setup

The hardware setup in this application is the same as in Section 4.7. The only difference here is that we send the estimated transfer function parameters instead of the input output data pair.

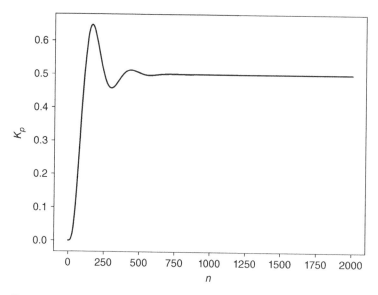

Figure 11.7 K_p value in time with MRAC setup.

11.5.2 Procedure

We will apply a random signal with amplitude 3 V, offset 7 V, and duration as 200 samples for one minute. Since the real-time RLS method limits the sampling frequency, we had to set it to 1 kHz. We use each sample to estimate the DC motor parameters. After one minute, we feed the estimated system parameter values to PC. We then compare the transfer function from estimated parameters and system identification (extracted in Chapter 6) in terms of time and frequency domain specifications. Due to computation cost and real-time realization requirements, we only formed C code version of the method next.

11.5.3 C Code for the System

For the application, we form a new project in Mbed Studio and use the C code in Listing 11.5. The software setup will be the same as in Section 3.4.3. As we execute this code, the RLS method is constructed using `RLS_Controller` class initialization function `RLS_Controller(unsigned int na, unsigned int nb, double gain, unsigned int cnt_max)`. Here, na and nb are the order of denumerator and numerator polynomials, respectively. `gain` is the RLS estimation gain. `cnt_max` is the number of samples to be used in estimation. After this value, estimation will be terminated and resulting transfer function will be accepted as final. In order to use the `RLS_Controller` class functions, the reader must go to `system_general.cpp` file and comment out the line `#define RLS_Disable` at the beginning of the file.

Listing 11.5: The C code to be used in implementing the RLS parameter estimation for the DC motor.

```
#include "mbed.h"
#include "system_general.h"
#include "system_input.h"
#include "system_output.h"
#include "system_controller.h"

Serial pc(USBTX, USBRX, 115200);
RawSerial device(PD_5, PD_6, 921600);
Ticker sampling_timer;
InterruptIn button(BUTTON1);
Timer debounce;
DigitalOut led(LED1);
PwmOut pwm1(PE_9);
PwmOut pwm2(PE_11);
DigitalOut pwm_enable(PF_15);
InterruptIn signal1(PA_0);
InterruptIn signal2(PB_9);
Timer encoder_timer;
AnalogIn ADCin(PA_3);
AnalogOut DACout(PA_4);
RLS_Controller rls(2, 2, 1000000, 5000);

int main()
{
    user_button_init();
    system_input_init(1, 12/3.3);
    system_output_init(1, 3591.84);
    sampling_timer_init();
    menu_init();
    while (true) {
        system_loop();
    }
}
```

If we check the `system_controller.cpp` file, there is an additional function `RLS_Controller::obtain_controller_output(float xnew, float ynew, int cnt)` to implement RLS algorithm sample by sample. This function takes the applied input data as `xnew`, obtained output data as `ynew`, and index of the signal as `cnt`. Then, it feeds 1 to output if `cnt` equals to preset number of the estimation process. Otherwise, it feeds −1. After the function feeds 1, system stops and waits for sending estimated transfer function coefficients to PC through UART communication. On the PC side, we receive the estimated transfer function coefficients by the Python code in Listing 11.6.

Listing 11.6: Obtaining estimated transfer function coefficients.

```
import mpcontrolPC
import mpcontrol

Ts = 0.00125
num, den = mpcontrolPC.get_coefficients_from_MC(3, 3, 'com10')
ge = mpcontrolPC.tf(num, den, Ts)
print(ge)
```

In Listing 11.6, we use the function get_coefficients_from_MC (NoNums, NoDens, com_port) to receive the estimated transfer function coefficients. Here, NoNums and NoDens stand for the number of numerator and denominator coefficients to be transferred, respectively. com_port is used to define which communication port is used on PC side for data transfer. As we obtain the numerator and denominator coefficients, we form the transfer function of the system.

11.5.4 Observing Outputs

We compare the transfer functions obtained by parameter estimation and system identification (extracted in Chapter 6) in this section. To do so, we first feed unit step signal, rectangular signal, and two sinusoidal signals with 2 and 100 Hz frequency to both systems. Then, we provide the output obtained from both systems. We also compare the time and frequency domain characteristics of the estimated and identified systems.

We provide comparison of the two transfer functions when step and rectangular signals are fed to them in Figure 11.8. The estimated system response is plotted by dashed line. The identified system response is plotted by a solid line. As can be seen in this figure, both systems have almost the same response to the selected input signals. The reader can also see this result by looking at Table 11.1.

We next provide the response of both estimated and identified systems for the two sinusoidal signals, with 2 and 100 Hz frequency, in Figure 11.9. Again, we use the dashed and solid lines for the estimated and identified system responses here. As can be seen in this figure, both systems provide almost the same output to the input sinusoidal signal with 2 Hz frequency. However, when the input sinusoidal signal with 100 Hz frequency is fed to both systems, the system constructed by the estimated parameters could not follow the identified system response. This difference can also be observed in Bode plot comparisons in Figure 11.10.

Noise in data originating from the quadrature encoder is the reason of the difference in Figures 11.9 and 11.10 for the 100 Hz sinusoidal input. The RLS method can find the low-frequency pole fairly well. However, the high-frequency pole cannot be estimated correctly due to noise. The reader can observe this by checking poles of the estimated and identified systems.

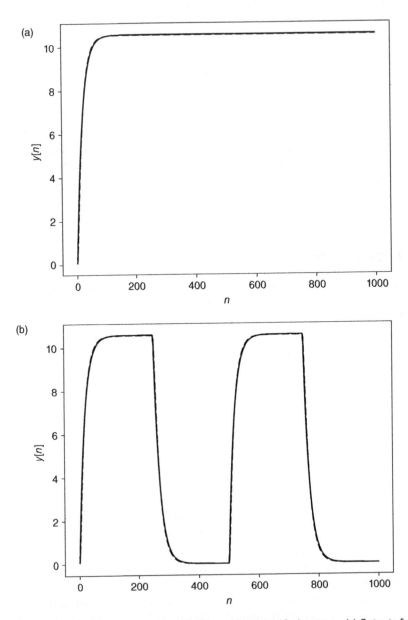

Figure 11.8 Output comparison of estimated and identified systems. (a) Outputs for step input. (b) Outputs for rectangular input.

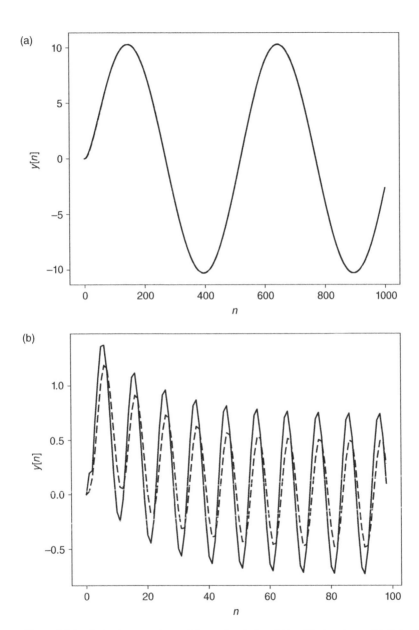

Figure 11.9 Output comparison of the estimated and identified systems. (a) Outputs for 2 Hz sinusoidal input. (b) Outputs for 100 Hz sinusoidal input.

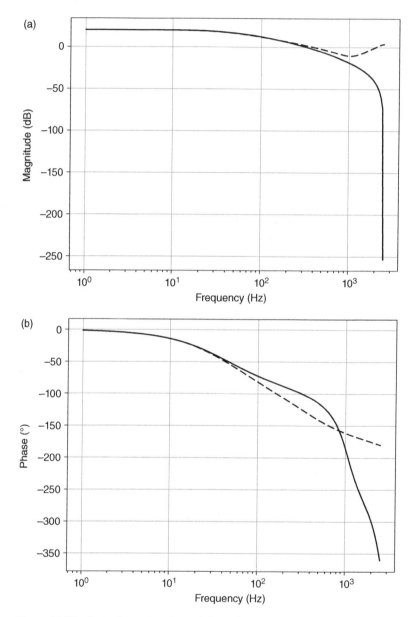

Figure 11.10 Bode plot comparison of the estimated and identified systems. (a) Magnitude plots. (b) Phase plots.

Table 11.1 Time domain performance criteria for the RLS estimated and system identified DC motor transfer functions.

Measure	RLS estimated	System identification
Rise time (T_{rt}), s	0.0513	0.0488
Settling time (T_{st}), s	0.0950	0.0888
Peak time (T_{pt}), s	0.4763	0.4375
Overshoot (M_p), %	0	0
Undershoot, %	0	0
Settling min.	9.5513	9.5326
Settling max.	10.5714	10.5536
Peak	10.5714	10.5536

11.6 Summary

This chapter focused on adaptive control schemes. We started with parameter estimation with RLS method. Then, we introduced ISTR and MRAC methods as indirect and direct controller design strategies. Finally, we applied the RLS method on estimating the parameters of the DC motor in real-time as the end of chapter application. Since we are using microcontrollers in implementation, adaptive control methods can be implemented on them easily. Therefore, the reader can benefit from them in solving real-life problems.

Problems

11.1 Consider the simple RC filter with transfer function

$$G(s) = \frac{1054}{s + 1055} \tag{11.19}$$

   a. represent this transfer function in digital form using bilinear transformation with $T_s = 0.0005$ seconds.
   b. apply RLS to estimate parameters (two numerator and two denominator parameters) of the digital transfer function.

11.2 Reconsider the simple RC filter in Problem 11.1,
   a. represent this transfer function in digital form using zero-order hold equivalent with $T_s = 0.0005$ seconds.

   b. form the ISTR with feedback controller for the digital transfer function. Select your own design criteria.
   c. form the ISTR with feedback and feedforward controllers for the digital transfer function for the design criteria in part (b).
   d. compare the closed-loop transient response measures of the two systems obtained in part (b) and (c).

11.3 Reconsider the simple RC filter in Problem 11.1,
   a. represent this transfer function in digital form using bilinear transformation with $T_s = 0.0005$ seconds.
   b. assuming the digital transfer function obtained in part (a) as $G(z)$, take the reference model as $G_m = 0.7G(z)$. Using MRAC, form a feedforward proportional controller for $G(z)$.

12

Advanced Applications

This chapter is divided into two parts as advanced controller design methods and digital control applications. In the first part, we will briefly introduce nonlinear, optimal, robust, and distributed control methods. We will also provide actual examples on these methods. In the second part, we aim to show how digital control concepts can be used to solve real-life problems. Therefore, we will provide several applications on open- and closed-loop control, visual servo, and control of a permanent magnet synchronous motor (PMSM). Through these applications, the reader will gain insight about the usage of classical and advanced digital control methods on real-life problems. Hence, we expect the book to serve its purpose.

12.1 Nonlinear Control

We assumed the system at hand or the controller to be designed as linear throughout the book. Unfortunately, this is not the case in practice. So, why did we follow such an approach? Well, linear system theory tools are easier to apply and approximate most control problems sufficiently well. Therefore, we followed the same path as with most control system books. There is also extensive research on nonlinear systems. Here, the analysis and design tools are different and provide very good results if applied properly. In this section, we start with nonlinear system identification by MATLAB followed by examples on nonlinear system control via Python control systems library.

12.1.1 Nonlinear System Identification by MATLAB

MATLAB can be used to model a nonlinear system. To do so, we should follow the steps mentioned in Chapter 6. However, we should select the "nonlinear models

Embedded Digital Control with Microcontrollers: Implementation with C and Python,
First Edition. Cem Ünsalan, Duygun E. Barkana, and H. Deniz Gürhan.
© 2021 The Institute of Electrical and Electronics Engineers, Inc. Published 2021 by John Wiley & Sons, Inc.
Companion website: www.wiley.com/go/Unsalan/Embedded_Digital_Control_with_Microcontrollers

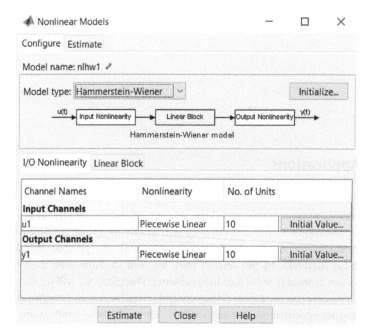

Figure 12.1 Nonlinear model properties window. (Source: Matlab, The MathWorks, Inc.)

..." option from the "Estimate →" dropdown menu. Then, a window opens up as in Figure 12.1. Here, we can enter the nonlinear model properties.

MATLAB provides different options for the model structure. Let's pick the Hammerstein–Wiener model as an example. This model has nonlinear elements at its input and output. Between these two, there is a linear system. As we feed the DC motor input and output data (obtained in Section 6.4) to MATLAB, we can estimate its Hammerstein–Wiener model. We set the pole and zero number as two, assign the delay value as one, and select input and output nonlinearities as piecewise linear functions for our example. The identification process results in 98.28% fit performance. We provide the modeling result in Figure 12.2. As can be seen in this figure, the model does not have input nonlinearity. On the other hand, there is a nonlinearity at output of the model. The linear system between the input and output has a fairly good step response.

Please note that, options provided by MATLAB in nonlinear system identification are plenty. We cannot cover all of them in this book. Therefore, the reader can consult the MATLAB help files for advanced usage of the system identification toolbox.

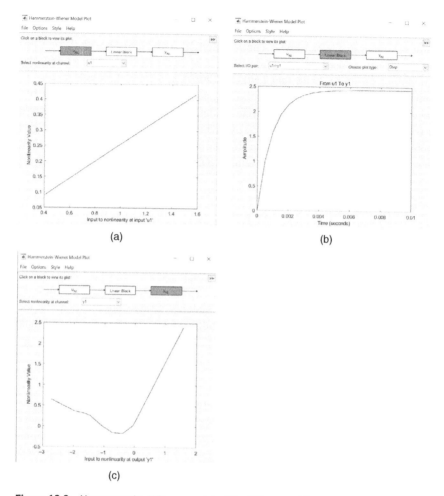

Figure 12.2 Hammerstein–Wiener model of the DC motor. (a) Input nonlinearity. (b) Output nonlinearity. (c) Step response of the linear system. (Source: Matlab, The MathWorks, Inc.)

12.1.2 Nonlinear System Input–Output Example

Python control systems library provides cruise control of a car example from the book by Åström and Murray (2008). Here, the authors used nonlinear model of a vehicle. They picked the PI controller and state space control for the controller design. The Python code for the example is named `cruise-control.py`, which can be found in the examples section of the Python control systems library website. The reader can gain insight about input and output properties of a nonlinear system and how it is controlled through this example.

12.1.3   Gain Scheduling Example

Gain scheduling is a nonlinear control method in which more than one linear controller is used in operation. Selection of these controllers is done by scheduling variables. Leith and Leithead (2010) provide a thorough review on gain-scheduling methods. Python control systems library provides a gain scheduling example based on a simple bicycle model. The Python code for the example is named `steering-gainsched.py`, which can be found in the examples section of the Python control systems library website. The reader can gain insight about the control of a nonlinear system via gain scheduling by this example.

12.1.4   Flat Systems Example

Flat system approach provides another perspective to design a controller for nonlinear systems. Please see the paper by Fliess et al. (1992) for details of this method. Python control systems library has an example on flat systems. In this example, trajectory of a nonlinear vehicle model while changing lanes is provided. The Python code for the example is named `kincar-flatsys.py`, which can be found in the examples section of the Python control systems library website. We direct the reader to this example for further understanding of flat system concepts and their implementation via Python.

12.1.5   Phase Portraits Example

Phase portrait provides trajectory of selected system variables. Although the method can be used to represent linear as well as nonlinear systems, we believe it is more helpful for the latter case. Hence, we included it in this section. Python control systems library provides several phase portrait examples from the book by Åström and Murray (2008). The Python code for these examples is named `phaseplots.py`, which can be found in the examples section of the Python control systems library website. Among these examples, we provide phase portrait of the inverted pendulum in Figure 12.3. This figure indicates that, the inverted pendulum has two attractors. Depending on starting point of the system, it converges to one of these attractors. For more information on the usage of phase portraits, please see Jordan and Smith (2007).

12.2   Optimal Control

Optimal control provides yet another perspective for the controller design. The aim here is designing a controller based on a given optimality criterion. Therefore, we should first define a cost function. Afterward, we can use mathematical

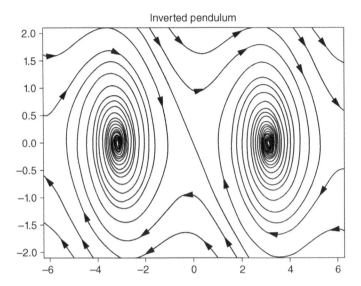

Figure 12.3 Phase portrait of the inverted pendulum.

tools to minimize the cost by taking into account the constraints and formalize the optimality criterion.

We consider the linear quadratic regulator (LQR) as the optimal control method in this section. To do so, we first introduce basic concepts. Then, we provide the Python control systems library example on continuous-time LQR control. Finally, we apply LQR to our DC motor.

12.2.1 The Linear Quadratic Regulator

LQR is a well-known method in optimal control. It computes the optimal gain matrix $\mathbf{K}$ such that the state-feedback law $\mathbf{u}[n] = -\mathbf{K}\mathbf{x}[n]$ minimizes the cost function

$$J = \sum_{n=1}^{\infty}(\mathbf{x}[n]^T\mathbf{Q}\mathbf{x}[n] + \mathbf{u}[n]^T\mathbf{R}\mathbf{u}[n] + 2\mathbf{x}[n]^T\mathbf{N}\mathbf{u}[n]) \qquad (12.1)$$

for the system with state-space equation $\mathbf{x}[n + 1] = \mathbf{F}\mathbf{x}[n] + \mathbf{G}\mathbf{u}[n]$. Here, $\mathbf{Q}$ and $\mathbf{R}$ are weight matrices. $\mathbf{N}$ is the cross weight matrix.

LQR finds the optimal feedback gain matrix $\mathbf{K}$ based on the solution $\mathbf{S}$ of the associated algebraic discrete-time Riccati equation, Bertsekas (2012),

$$\mathbf{F}^T\mathbf{S}\mathbf{F} - \mathbf{S} - (\mathbf{F}^T\mathbf{S}\mathbf{G} + \mathbf{N})(\mathbf{G}^T\mathbf{S}\mathbf{G} + \mathbf{R})^{-1}(\mathbf{G}^T\mathbf{S}\mathbf{F} + \mathbf{N}^T) + Q = 0 \qquad (12.2)$$

Hence, we will have

$$K = (G^T S G + R)^{-1}(G^T S F + N^T)$$ (12.3)

We can use this gain value in the regulator structure given in Section 10.1.2. Hence, the LQR is formed.

12.2.2 Continuous-Time LQR Example

Python control systems library has an example on the usage of continuous-time LQR. This example takes a planar vertical takeoff and landing aircraft system to be controlled as provided in Åström and Murray (2008). The Python code for the example is named `pvtol-lqr.py`, which can be found in the examples section of the Python control systems library. The reader can gain insight on the principles of continuous-time optimal control design strategies via this example.

12.2.3 LQR for the DC Motor

We next provide an actual example on the usage of the digital LQR on our DC motor. To do so, we benefit from the Python code in http://www.mwm.im/lqr-controllers-with-python/. The author in this website emphasizes that the code originates from the book by Bertsekas (2012). We provide the modified Python code for this example in Listing 12.1.

Listing 12.1: LQR code for the DC motor.

```
import numpy as np
import scipy.linalg

def dlqr(A,B,Q,R):
        X = np.matrix(scipy.linalg.solve_discrete_are(A, B, Q, R))
        K = np.matrix(scipy.linalg.inv(B.T*X*B+R)*(B.T*X*A))
        return K

F = [[1.8188, -0.8224], [1.0, 0.0]]
G = [[1.0], [0.0]]
C = [[0.036, 0.0017]]

Fn = np.array(F)
Gn = np.array(G)
Cn = np.array(C)

Q = 1.1*np.transpose(Cn)*Cn
R = 1
K=dlqr(Fn,Gn,Q,R)

print(K)
```

As we execute the Python code in Listing 12.1, we obtain the gain matrix as K=[[0.1464 -0.1131]]. We can use this gain matrix in the regulator code for the DC motor in Listing 10.2. Hence, we obtain the output and state-space plots as in Figures 12.4(a) and 12.4(b). In order to compare these results with the previously introduced pole placement method in regulator design, we provide

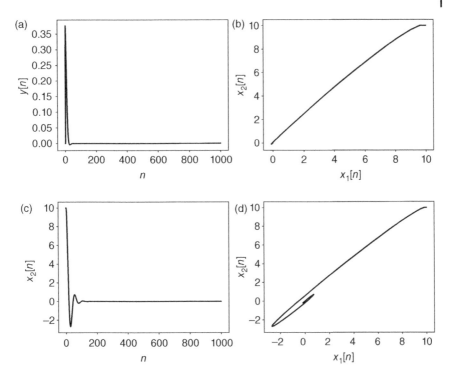

Figure 12.4 Comparison of the regulators designed by pole placement and LQR. (a) Output, LQR. (b) State space, LQR. (c) Output, pole placement. (d) State space, pole placement.

these results in Figures 12.4(c) and 12.4(d). As can be seen in these figures, LQR leads the system output to zero faster than the regulator designed by the pole placement method.

12.3 Robust Control

The aim in robust control is to design the controller such that it can deal with system uncertainties. To do so, we assume that some system parameters are unknown but bounded. The reader can check the book by Zhou and Doyle (1997) for further understanding robust control concepts.

Python control systems library provides two examples on the usage of robust control based on the book by Skogestad and Postlethwaite (1996). As usual, these can be found in the examples section of the library website. The Python code for the first example, on a SISO system, is named `robust-siso.py`. The Python code for the second example, on a MIMO system, is named `robust-mimo.py`.

We direct the reader to these examples to grasp the usage of robust control methods via Python control systems library.

12.4 Distributed Control

Distributed (or networked) control deals with control problems in which more than one system or controller is in different locations. These concepts are very important in today's world as devices become connected and internet of things (IoT) becoming a reality. The reader can check the book by Yu et al. (2016) for distributed control concepts. To note here, there are also other valuable resources on this topic.

In this section, we provide a simple but practical distributed control application. The aim here is to observe and control speed of the DC motor from a smart phone. To do so, we will use a bluetooth module connected to the STM32 board controlling the DC motor. We will also develop an application on the smart phone to control and display motor speed on its screen.

12.4.1 Hardware and Software Setup

Schematic representation of the hardware and software setup for this application is given in Figure 12.5. The software setup is formed by the DC motor control

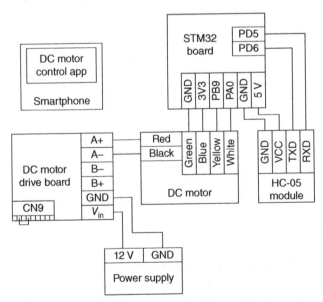

Figure 12.5 Hardware and software setup for the distributed control application.

application on the smart phone. The hardware setup consists of DC motor, STM32 board, motor driver board, and HC-05 bluetooth module.

The smart phone application can be constructed by the MIT App Inventor program which runs on the browser. It has a user friendly graphical user interface (GUI). The reader can find several bluetooth applications prepared by the MIT App Inventor. These may be of help in preparing the software setup for DC motor control.

The reader can use the HC-05 bluetooth module connected to the STM32 board on the hardware side. Hence, data transferred from the smart phone via bluetooth communication can be acquired on the microcontroller side. Although the HC-05 module is based on the early bluetooth 2.0 technology, it is still widely used and provides fairly good results in operation.

12.4.2 Procedure

The first step in the application is to set up the HC-05 module via UART communication. To do so, we should use the FT232 module. Initially, we should connect the RXD, TXD, GND, and VCC pins of HC-05 to TX, RX, GND, and VCC pins of FT232, in the same order. Then, we should connect the FT232 module to PC and open a terminal program (such as Tera Term). Baud rate in the terminal program should be set as 9600 bps since this is the default value for the HC-05 module. Afterward, we should use AT commands on the terminal for setting up the module. For more information on AT commands, please see Ünsalan et al. (2017).

In order to setup the HC-05 module, let's start with entering the first command AT through the terminal program. If the hardware is set correctly, the module should return OK to the terminal. Next, we should assign a name to the HC-05 module. This will allow us to reach it from smart phone. To do so, the command AT+NAME=DesiredName should be entered through the terminal. The HC-05 module should return OK. Finally, we should set the baud rate for UART communication. Therefore, we should use the command AT+UART=921600,0,0 on the terminal. This command has the setup such that 921600 stands for the baud rate, first 0 represents the stop bit, and second 0 represents the parity bit in UART communication. Again, the HC-05 module should return OK to the terminal. Afterward, the HC-05 module will be ready to be used. To note here, the HC-05 module will have its baud rate set as 921600 bps from this point on. This value should be used for all UART communication operations between the module and another device. The reader can reach more information on this module and other wireless communication modules from our book (Ünsalan et al. 2017).

The next step in the procedure is to form the application on smart phone. The application should be set up such that it should locate nearby bluetooth devices. It should allow connecting to a selected device. As the connection is established, speed of the DC motor should be modified, between 0 and maximum allowed value, by a slider in the application. There should also be another button to change rotation direction of the motor. Finally, there should be a window in the application continuously displaying speed of the DC motor. To note here, the speed value will be read from encoder of the motor and transmitted back to smart phone from the STM32 board. There should also be a button in the application to stop bluetooth connection.

We should have all necessary software running on the STM32 microcontroller to control the DC motor. Here, we suggest forming a closed-loop controller to set speed of the microcontroller. All these mentioned steps provide insight on the basics of distributed control. Moreover, the methodology introduced in this application can be used to form more complex distributed control applications such as connecting more than one controller or system via bluetooth communication.

12.5 Auto Dimmer

This application aims to form an open-loop controller to adjust brightness of an LED based on ambient light in the environment. The controller will decrease brightness of the LED if ambient light is high. It will increase brightness of the LED if ambient light is low.

12.5.1 Hardware Setup

We provide schematic representation of the hardware setup for this application in Figure 12.6. In this setup, light dependent resistor (LDR) is used to measure ambient light in the environment. LDR is a resistive element with its resistance

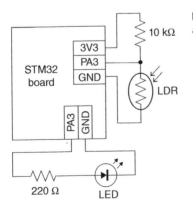

Figure 12.6 Hardware setup for the auto dimmer application.

changing by light intensity falling on it. If the light intensity is high, the LDR resistance falls through 0 Ω. For low light, the LDR resistance increases.

The LDR can be serially connected to another resistor as in Figure 12.6 to form a voltage divider. Hence, the formed setup provides a voltage between 0 and 3.3 V to ADC pin (PA3 here) of the microcontroller. We can use DAC module on the microcontroller with its pin PA4 to adjust brightness of the LED.

12.5.2 Procedure

The first step in the procedure of the application is to calibrate the overall system. Hence, we will set the minimum and maximum ambient light levels to turn on and off the LED. Within these two levels, LED brightness will change linearly based on measured voltage from the LDR. This way, we form an open-loop controller with gain parameter (or a proportional controller). In fact, the formed system can also be taken as nonlinear since we set minimum and maximum voltage levels in operation. Finally, we can replace the LED with an adjustable light bulb to implement a more general dimmer.

12.6 Constructing a Servo Motor from DC Motor

A servo motor is actually a DC motor having internal feedback loop. Servo motors are used for rotation control between 0° and 180°. Hence, the servo motor has a gearbox to change the high speed low torque property of the motor to low speed high torque. Afterward, an internal feedback loop is formed.

The feedback signal is generated by connecting a linear potentiometer to output of the gearbox in simple servo motors. Resistance value obtained from the potentiometer for rotation angles between 0° and 180° is assumed to be known. This way, the motor is rotated till the required angle is reached by looking at the resistance value. Advanced servo motors have feedback loop formed by a connected encoder having index pin. The feedback signal is formed by the encoder for this case. In this application, we will construct a servo motor from DC motor.

12.6.1 Hardware Setup

We will construct the servo motor from the DC motor in two different ways. In the first setup, we will pick a simple 5 V DC motor having no encoder. We will attach a 10 kΩ potentiometer to shaft of the motor. Then, we will form a closed-loop controller. The setup for this approach is given in Figure 12.7.

In the second setup, we will pick the Pololu DC motor used in previous chapters. Since this motor has a gearbox, it can provide low speed and high torque

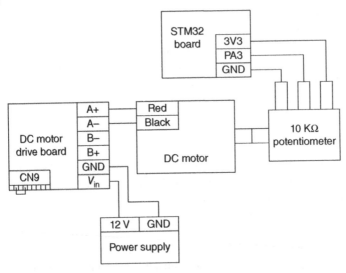

Figure 12.7 Hardware setup for the servo motor from DC motor application using potentiometer feedback.

as required from the servo motor. The encoder of this motor does not have an index pin. Therefore, we will use the ADXL-345 digital accelerometer (attached to shaft of the motor) as the feedback module. Then, we will form the closed-loop controller. We provide the setup for this approach in Figure 12.8.

12.6.2 Procedure

In the first part of the application, resistance of the potentiometer for 0° and 180° is obtained. Resistance of the potentiometer changes linearly with its rotation angle. Hence, it becomes possible to obtain angle of the motor shaft from resistance of the attached potentiometer. To do so, we should form a voltage divider circuit by the potentiometer and read the voltage from the ADC module of the STM32 microcontroller. This voltage value can be used as the feedback signal to form a closed-loop control action. Since the overall system is formed by the DC motor and potentiometer, we can form the transfer function between the required angle (as reference signal) and motor rotation angle (as output signal). This way, the reader can design a controller for this setup.

In the second part of the application, we should know how to get feedback signal from the accelerometer attached to shaft of the DC motor. To do so, we will use the y and z axis of the accelerometer. Assume that the accelerometer attached to the motor shaft is parallel to ground. For this case, the accelerometer will give $-g$ (gravity) and 0 values from its z and y axis, respectively. As we rotate the accelerometer

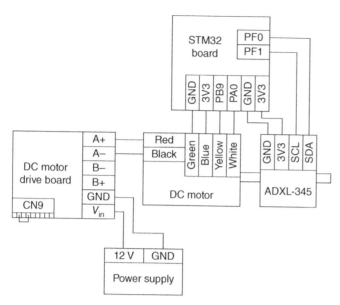

Figure 12.8 Hardware setup for the servo motor from DC motor application using accelerometer feedback.

through 90° with the motor shaft, the value read from the z axis will be close to 0. Meanwhile, the value read from the y axis will approach g. At exactly 90°, these values will be 0 and g, respectively. From 90° to 180°, the z axis values will approach g. Meanwhile, the y axis value will approach 0. At exactly 180°, these values will be g and 0, respectively. By looking at the sign of the z axis data, we can decide whether the motor rotation angle is between 0° and 90° or 90° and 180°. Then, we can use y axis data to calculate the actual angle. Based on these and required rotation angle, we can form an closed-loop controller. Hence, we can rotate the DC motor based on the reference input and feedback signal from the accelerometer.

12.7 Visual Servoing

The term visual servoing indicates a vision-based control operation. Here, the vision sensor (camera most of the times) is used as the feedback module. Then, a closed-loop controller is formed to control a system (such as robot arm). In order to show how visual servoing can be constructed, we will set position of the DC motor based on the acquired image from a camera in this application. Hence, the reader can gain insight on visual servoing operations.

12.7.1 Hardware Setup

The hardware setup for the visual servoing system to be constructed is in Figure 12.9. As can be seen in this figure, the OV7670 camera module is used to acquire image. The camera is connected to DCMI module of the microcontroller. This module is used to acquire camera image in a synchronous manner through its HREF, VREF, and PCLK signals. Image data is obtained from the camera in parallel manner through eight data bits of the DCMI module. The XCLK pin of the OV7670 camera is connected to master clock output pin of the STM32 microcontroller. Hence, the necessary clock signal for the camera is provided by the microcontroller.

12.7.2 Procedure

A green ball is moved within camera's point of view in the application. First, we acquire image from the camera and locate the ball by simple image processing

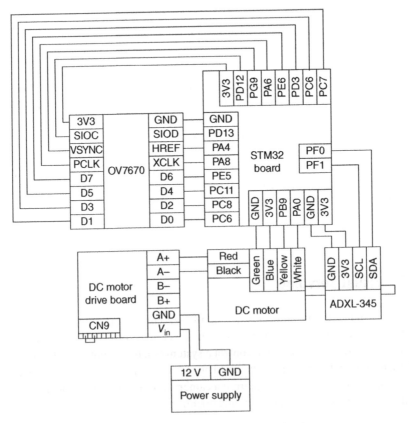

Figure 12.9 Hardware setup for the visual servoing application.

techniques. Then, we rotate the DC motor accordingly. If the ball is at the center of the image, then we set the DC motor rotation angle to 0°. If the ball is at the leftmost or rightmost location, then the rotation angle becomes −90° and 90°, respectively. For the ball location between these two values, we set the rotation angle linearly. In order to measure rotation angle of the motor, we use the ADXL-345 accelerometer as in the servomotor application.

12.8 Smart Balance Hoverboard

Smart balance hoverboard is a two wheeled self-balancing vehicle in which the user stands on it and moves in any direction. This is an unstable system such that the hoverboard tries to balance itself via several gyroscopes and accelerometers. The vehicle senses body position of the user though these sensors. The closed loop formed within the vehicle allows balancing the vehicle by rotating its wheels. In this application, we will form a simplified hoverboard system.

12.8.1 Hardware Setup

We provide setup for the smart balance hoverboard application in Figure 12.10. As can be seen in this figure, we will balance the DC motor shaft by the ADXL-345 accelerometer attached to it. Hence, a simplified hoverboard is formed.

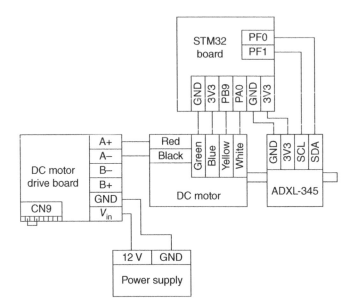

Figure 12.10 Hardware setup for the smart balance hoverboard application.

12.8.2 Procedure

The aim in this application is keeping y axis of the accelerometer (attached to shaft of the DC motor) parallel to ground under external forces. In other words, when we move the motor by hand to left or right, it will react by rotating its shaft to keep the accelerometer parallel to ground. For this purpose, we will acquire angle data from the accelerometer as in the servomotor application. The accelerometer's parallel location to ground will be represented by 0°. When we move the motor around its y axis, this angle will have positive or negative values. We will form a closed-loop control to set the angle to 0° every time when such a movement occurs.

12.9 Line Following Robot

Line following robot is a popular application showing how to navigate a robot via drawn line on the ground. In this application, we will form such a system by the Zumo robot chassis manufactured by Pololu. The system has two DC motors to move the robot. It also has a sensor module to measure light intensity reflected from the ground. We can represent the overall system as a closed-loop controller such that the line to be followed is the reference input signal. The light sensor module forms the feedback loop. Based on the reference input and feedback, the motors keep the robot follow the line.

12.9.1 Hardware Setup

The hardware setup for the line following robot application is as in Figure 12.11. All the equipment given here will be placed on the Zumo robot chassis. The reader can check the link https://www.pololu.com/product/1418 to see how this can be done.

In this application, power to the STM32 board is provided by four AA batteries. Hence, the JP1 jumper on the board should be disconnected. Afterward, the JP3 jumper should be disconnected from its current location and connected to the leftmost E5V pin. Then, positive and negative sides of the battery pack should be connected to E5V and any ground pin of the STM32 board, respectively.

12.9.2 Procedure

The aim in this application is letting the Zumo robot to follow a line drawn on ground. To do so, the QTRX-HD-04RC reflectance sensor array is used. This sensor has four IR LED/phototransistor pairs. In each pair, the IR LED submits light and phototransistor acquires the reflected light from the target surface. For

Figure 12.11 Hardware setup for the line following robot application.

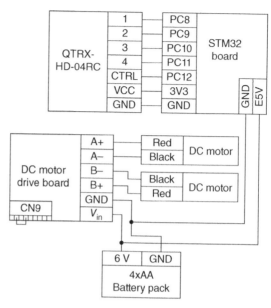

our case, the target surface is a black line on white background. This way, the QTRX-HD-04RC reflectance sensor array can be used to determine location of the robot relative to the black line to be followed. Then, motor speeds are adjusted to keep the robot follow the line. Hence, a closed-loop controller is formed. To note here, the camera setup in the visual servo application can also be used for this application instead of the QTRX-HD-04RC reflectance sensor array.

12.10 Active Noise Cancellation

Active noise cancellation (ANC) is used to eliminate unwanted noise on a signal. Therefore, it is used in several applications. In this application, we will simulate the ANC usage by a simple example. To do so, we will pick the least means square (LMS) method explained in detail in the book by Kuo and Morgan (1996).

12.10.1 Hardware Setup

The signal to be used in the ANC operation will be obtained from an external signal generator. Based on this, hardware setup to be used in the application will be as in Figure 12.12.

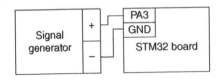

Figure 12.12 Hardware setup for the active noise cancellation application.

12.10.2 Procedure

General setup for ANC is given in Figure 12.13. As can be seen here, the noise term is obtained by a sensor. The ANC actuator is used to generate the signal to eliminate unwanted noise. The error sensor forms a feedback loop to control whether the noise is eliminated or not. Noise cancellation operation is performed in the ANC algorithm module. In this application, we will use the most popular method in literature called LMS within the module.

Block diagram of the ANC operation by the LMS algorithm is as in Figure 12.14. Here, $s[n]$ represents the original signal not affected by noise. We will generate this signal from the signal generator. $x[n]$ represents the noise detected by the sensor. We will generate this signal by the STM32 board. $x'[n]$ represents the actual noise term affecting the original signal. Therefore, we will cancel this noise signal. This signal will be generated within the STM32 board by filtering $x[n]$ with the filter $P(z)$. In other words, the $P(z)$ filter represents the path affecting the noise from its source to reaching the original signal. In this application, we will represent it by

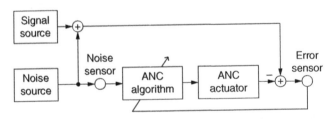

Figure 12.13 General representation of the ANC system.

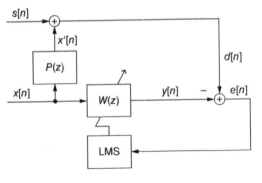

Figure 12.14 Block diagram of LMS based ANC operation.

an FIR filter. In Figure 12.14, $W(z)$ is the adaptive FIR filter which will generate the noise cancellation signal. Coefficients of this filter will be obtained by the LMS algorithm. $d[n]$ represents the original signal with noise added on it. Finally, $e[n]$ is the feedback signal to be sent to the LMS algorithm. This signal is obtained by the difference of signals $d[n]$ and $y[n]$.

We pick working frequency for this application as 10 kHz. Therefore, the system will acquire data from the signal generator; generate noise sample; feed it to $P(z)$ to obtain the error term affecting the original signal; and run the LMS algorithm to find the $W(z)$, adaptive FIR filter, coefficients at every 0.1 ms. We pick the input signal as sinusoidal with amplitude between 1 and 2 V. The reader can select the signal frequency as desired based on the sampling frequency. As the system is executed, working principles of ANC can be observed through simulation.

12.11   Sun Tracking Solar Panel

Solar energy has become an important green energy source recently. To do so, solar panels are used to harvest energy from the sunlight. Unfortunately, these panels have low efficiency. Therefore, they should be positioned such that they always see the sun as good as possible. The sun tracking solar panel aims to achieve this by rotating the solar panel based on the sun location. We will form such a system in this application in two different ways. First, we will use a stepper motor and form an open-loop controller accordingly. Second, we will use the DC motor and form a closed-loop controller for this purpose.

12.11.1   Hardware Setup

We provide system setup formed by the stepper motor in Figure 12.15. The stepper motor rotates in steps for the given voltage pattern to its pins. Therefore, it does not have a continuous movement as in the DC motor. Full rotation step number is known for the stepper motor. Therefore, the required rotation angle of the motor shaft can be set by the corresponding step number. Speed in the rotation operation is set by the duration between steps. For more information on the usage of stepper motor please see Ünsalan and Gürhan (2013). In this application, we will use the stepper motor having 400 steps for full rotation. We provide the DC motor based closed-loop setup for the sun tracking solar panel in Figure 12.16. Here, we have a similar setup for our DC motor as in previous chapters.

12.11.2   Procedure

As can be seen in Figures 12.15 and 12.16, there are two LDRs in the system. These should be placed on far ends of the solar panel. Hence, the light intensity received

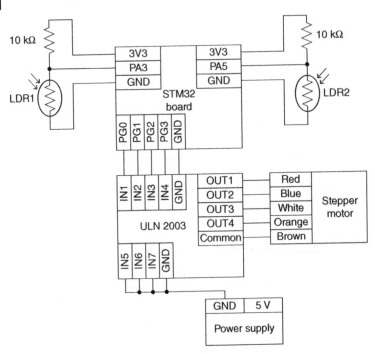

Figure 12.15 Hardware setup for the sun tracking solar panel application using stepper motor.

by these two LDRs can guide the system. When the sun is on top, we expect the two LDRs receive the same light intensity. If one LDR receives more light compared to the other, this means that the solar panel is not directly looking at the sun. Therefore, it should be rotated to have the same reading from two LDRs. This way, it is assured that the panel gets the maximum sunlight. When the stepper motor is used, the rotation angle to be moved to is converted to the corresponding step number. Then, the open-loop control action moves the panel to that angle. For the DC motor case, angle of the motor is measured by the ADXL-345 accelerometer attached to shaft of the motor. This way, we can form a feedback loop such that the motor rotating the solar panel performs the closed-loop control action.

12.12 System Identification of a Speaker

This application aims to obtain transfer function of a speaker. To do so, we will use the system identification tools introduced in Chapter 6.

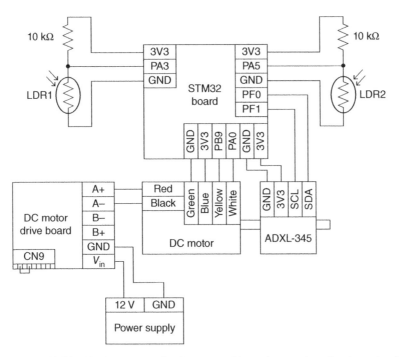

Figure 12.16 Hardware setup for the sun tracking solar panel application using DC motor.

12.12.1 Hardware Setup

We provide hardware setup for the system to be used in this application in Figure 12.17. Here, PAM-8403 is a class-D audio amplifier which can provide 3 W output without needing any external filter. The ADXL-345 accelerometer should be located on the speaker to measure its cone displacement. To do so, we should take double integration of the acceleration data.

12.12.2 Procedure

When a voltage is applied across terminals of the speaker the cone inside the speaker moves in and out causing pressure waves perceived as sound. There are two parts inside the speaker as mechanical and electrical. These are shown in Figure 12.18.

As can be seen in Figure 12.18, the speaker consists of a fixed magnet that produces uniform magnetic field of strength β (the power factor). The speaker has a cone with mass (M_{ms}) that moves in x direction. The cone can be modeled by a spring (K_{ms}) and friction (B_{ms}) element. There is a coil of wire with radius a within

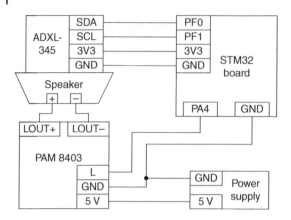

Figure 12.17 Hardware setup for the system identification of a speaker application.

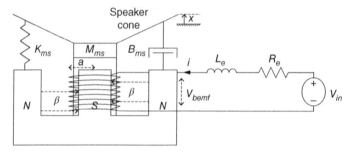

Figure 12.18 Mechanical and electrical parts of the speaker.

the magnetic field and it is attached to the cone. The wire has resistance (R_e) and inductance (L_e). The coil consists of n turns and it moves along with the cone. A force is generated that moves the cone when current passes through the coil. Back emf voltage is also generated in the coil when the cone moves. V_{bemf} and V_{in} are the back emf and input voltages, respectively. Speed and displacement of the cone is represented as v and x, respectively. Based on these definitions, transfer function of the speaker between the input, $V_{in}(s)$, and output, $X(s)$ is obtained by Palm (2013) as

$$\frac{X(s)}{V_{in}(s)} = \frac{\beta}{(L_eM_{ms})s^3 + (R_eM_{ms})s^2 + (L_eK_{ms} + \beta^2)s + R_eK_{ms}} \quad (12.4)$$

Unfortunately, most speaker parameters are not tabulated in datasheets. Moreover, experimental setup to obtain these parameters is not simple. Therefore, the most suitable method to obtain transfer function of a speaker is using system identification tools. The mathematical derivation provided above indicates that the system should be third order. Using this information, the reader can form the identification setup. The input should be a random signal for time domain system identification. As the acceleration data is obtained, it should be filtered before taking double integration to obtain the displacement information. This will be the output signal obtained from the speaker. Hence, this input output signal pair can be used in the system identification setup.

12.13 Peltier Based Water Cooler

Peltier is a device which can heat up its one side and cool down its other side based on voltage applied to it. We will construct a peltier based water cooler in this application.

12.13.1 Hardware Setup

We provide hardware setup for this application in Figure 12.19. In this figure, DS18B20 is the waterproof temperature sensor which can work between $-10°$ and $85 °C$ with $0.5°$ precision. For more information on the working principles of this sensor, please see its datasheet. TEC1 12703 is the peltier module with 12 V and 3 A rating. When constant current is applied to the module, its one side heats up and the other side cools down. After some time, this temperature difference vanishes. Therefore, a heat sink is added to the heating up side of the peltier to decrease the temperature further on its cooler side in operation. The IRL540N MOSFET transistor is used to control the current required by the peltier. This transistor is controlled by DAC module of the STM32 microcontroller.

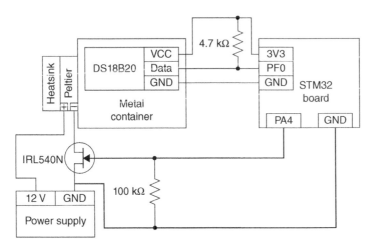

Figure 12.19 Hardware setup for the peltier based water cooler application.

12.13.2 Procedure

In this application, we will use the peltier and the temperature sensor to cool down the water in a metal container to a desired level. The DS18B20 temperature sensor forms the feedback signal. We will use this signal and a PI controller to construct a closed-loop controller for the peltier module. The reader should apply a step signal and store temperature values till the peltier temperature stabilizes. Then, these values should be used in the system identification framework introduced in Chapter 6 to form the transfer function of the peltier. Here, a first-order system will be sufficient to model the peltier. Then, the obtained transfer function can be used to design the PI controller in the closed-loop operation.

12.14 Controlling a Permanent Magnet Synchronous Motor

The permanent magnet synchronous motor (PMSM) has the characteristics of both brushless DC and induction motors. It has a permanent magnet rotor as in the brushless DC motor and stator structure generating sinusoidal back emf as in the induction motor. The PMSM needs full torque control at zero speed, instant speeding and stopping, and smooth rotation at high speed. To achieve these goals, field-oriented control (FOC) is used. This method can be explained in simple terms as dividing the stator current into two parts as magnetic field and torque producing. Then, these parts can be controlled separately.

ST Microelectronics has a firmware called X-CUBE-MCSDK to control PMSM. This firmware consists of the STM32 motor control workbench program and FOC library. The reader can download this firmware from https://www.st.com/en/embedded-software/x-cube-mcsdk.html.

12.14.1 Hardware Setup

The X-CUBE-MCSDK firmware allows using specific STM32 development board paired with selected motor driver board. Unfortunately, the firmware does not handle the STM32 board used throughout the book. Therefore, we picked the NUCLEO-F303RE development board with X-NUCLEO-IHM16M1 motor driver board for this application. To use this pair, the reader should place the motor driver board onto the NUCLEO-F303RE development board thorough its header pins. Then, the motor driver board should be connected to a 12 V adapter. Finally, the reader should connect phase pins of the PMSM to U, V, and W inputs of the motor driver. We provide the hardware setup for this application in Figure 12.20.

Figure 12.20 Hardware setup for controlling the PMSM.

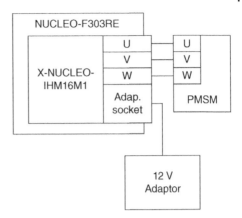

The reader can select either the iPower GBM2804H-100T or Bull-Running BLDC model BR2804-1700 motor for this application. The former has sockets on it. Hence, it can be directly connected to the motor driver board. The reader should use a screwdriver to connect the Bull-Running BLDC motor. Besides, another motor can also be selected by the reader as long as it satisfies the voltage and maximum current values.

12.14.2 Procedure

As we start the STM32 motor control workbench program, its main window will be as in Figure 12.21. The FOC library needs model of the PMSM to generate necessary control parameters. To do so, we will need the motor profiler program which can be reached from the top right of the main window.

The opened motor profiler window will be as in Figure 12.22. Here, we should first select the development board and motor driver pair to be used through the "Select Boards" section. Then, we should press "Connect" to connect the boards to the motor profiler. Afterward, we should enter the PMSM pole pair number, maximum speed, maximum current, and applied voltage to "Pole Pairs," "Max Speed," "Max Current," and "VBus" boxes, respectively. Finally, we should select the PMSM type as either "SM-PMSM" or "I-PMSM." Afterward, we can start the PMSM modeling operation by pressing "Start Profile." The motor profiler applies different input signals to the PMSM and identifies its parameters. As the modeling finalizes, we should save the result by pressing on "Save" and exit the motor profiler window.

Next, we should open a new project under STM32 motor control workbench by pressing on "New Project." A new window should open up as in Figure 12.23.

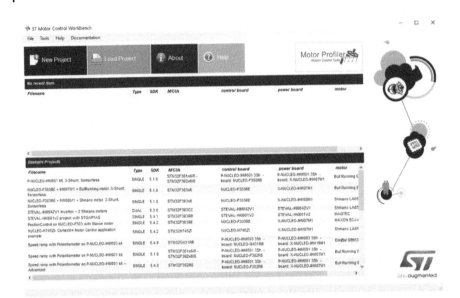

Figure 12.21 STM32 motor control workbench program, main window. (Source: STMicroelectronics.)

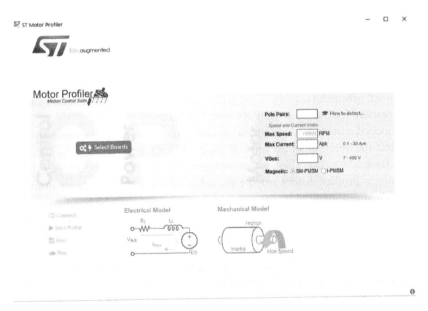

Figure 12.22 Motor profiler window. (Source: STMicroelectronics.)

Figure 12.23 New project window. (Source: STMicroelectronics.)

Here, we should select the development board from the "Control" drop-down menu; motor driver from the "Power" menu; and saved motor model from the "Motor" menu.

As we press "OK" to proceed, this leads us to STM32 motor control workbench GUI given in Figure 12.24. The reader can adjust and modify the PSMS control setup from this GUI. Moreover, the reader can press the "Project Generation" section (labeled by a light gray arrow on top of the screen) to form a project to be used for selected development environments such as STM TrueStudio, Keil, or IAR.

The STM32 motor control workbench also allows controlling the PMSM over its GUI. To do so, we should press "Monitor" on top of the GUI to open the monitor window as in Figure 12.25. Here, we should first connect our system by pressing

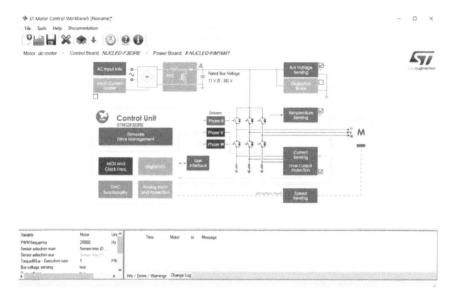

Figure 12.24 STM32 motor control workbench GUI. (Source: STMicroelectronics.)

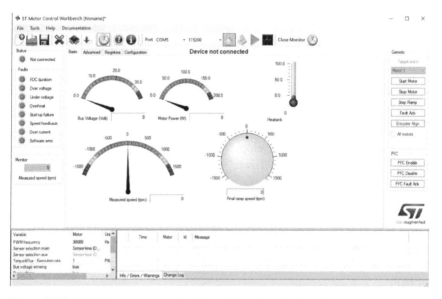

Figure 12.25 STM32 motor control workbench monitor window. (Source: STMicroelectronics.)

"Connect." Then, we should start PMSM by pressing on "Start Motor." We can adjust the speed applied to the motor by turning a knob on the GUI. We can observe the actual speed value after applying FOC from the "Measured Speed" section. If we want to stop the motor, we should press the "Stop Motor" button.

Appendix A

STM32 Board Pin Usage Tables

Pins on the STM32 board can be used for various purposes. Usage areas of each pin are summarized in Tables A.1–A.8. In these tables, we only summarized the usage areas to be considered in this book. Other usage areas of the mentioned pins can be found in https://os.mbed.com/platforms/ST-Nucleo-F767ZI/.

Embedded Digital Control with Microcontrollers: Implementation with C and Python,
First Edition. Cem Ünsalan, Duygun E. Barkana, and H. Deniz Gürhan.
© 2021 The Institute of Electrical and Electronics Engineers, Inc. Published 2021 by John Wiley & Sons, Inc.
Companion website: www.wiley.com/go/Unsalan/Embedded_Digital_Control_with_Microcontrollers

Table A.1 Pin usage table for the STM32 board, connector CN7-left.

Pin	Port name	Usage area
1	PC6	Digital I/O, Interrupt, Timer, UART
2	PB15	Digital I/O, Interrupt, Timer, UART, SPI
3	PB13	Digital I/O, Interrupt, Timer, UART, SPI
4	PB12	Digital I/O, Interrupt, Timer, UART, SPI
5	PA15	Digital I/O, Interrupt, Timer, UART, SPI
6	PC7	Digital I/O, Interrupt, Timer, UART
7	PB5	Digital I/O, Interrupt, Timer, UART, SPI, I^2C
8	PB3	Digital I/O, Interrupt, Timer, UART, SPI
9	PA4	Digital I/O, Interrupt, UART, SPI, ADC, DAC
10	PB4	Digital I/O, Interrupt, Timer, UART, SPI

Table A.2 Pin usage table for the STM32 board, connector CN7-right.

Pin	Port name	Usage area
1	PB8	Digital I/O, Interrupt, Timer, UART, I^2C
2	PB9	Digital I/O, Interrupt, Timer, UART, I^2C
3	AVDD	External analog voltage/voltage reference pin
4	GND	Ground voltage
5	PA5	Digital I/O, Interrupt, Timer, SPI, ADC, DAC
6	PA6	Digital I/O, Interrupt, Timer, SPI, ADC
7	PA7	Digital I/O, Interrupt, Timer, SPI, ADC
8	PD14	Digital I/O, Interrupt, Timer, UART
9	PD15	Digital I/O, Interrupt, Timer, UART
10	PF12	Digital I/O, Interrupt

Table A.3 Pin usage table for the STM32 board, connector CN8-left.

Pin	Port name	Usage area
1	NC	–
2	IOREF	I/O reference
3	RESET	External reset
4	+3V3	3.3 V input/output
5	+5V	5 V output
6	GND	Ground voltage
7	GND	Ground voltage
8	VIN	External 7–12 V input

Table A.4 Pin usage table for the STM32 board, connector CN8-right.

Pin	Port name	Usage area
1	PC8	Digital I/O, Interrupt, Timer, UART
2	PC9	Digital I/O, Interrupt, Timer, UART, I²C
3	PC10	Digital I/O, Interrupt, UART, SPI
4	PC11	Digital I/O, Interrupt, UART, SPI
5	PC12	Digital I/O, Interrupt, UART, SPI
6	PD2	Digital I/O, Interrupt, Timer, UART
7	PG2	Digital I/O, Interrupt
8	PG3	Digital I/O, Interrupt

Table A.5 Pin usage table for the STM32 board, connector CN9-left.

Pin	Port name	Usage area
1	PA3	Digital I/O, Interrupt, Timer, UART, ADC
2	PC0	Digital I/O, Interrupt, ADC
3	PC3	Digital I/O, Interrupt, SPI, ADC
4	PF3	Digital I/O, Interrupt, ADC
5	PF5	Digital I/O, Interrupt, ADC
6	PF10	Digital I/O, Interrupt, ADC
7	NC	—
8	PA7	Digital I/O, Interrupt, Timer, SPI, ADC
9	PF2	Digital I/O, Interrupt, I^2C
10	PH1	Digital I/O, Interrupt
11	PH0	Digital I/O, Interrupt
12	GND	Ground voltage
13	PD0	Digital I/O, Interrupt, UART
14	PD1	Digital I/O, Interrupt, UART
15	PG0	Digital I/O, Interrupt

Table A.6 Pin usage table for the STM32 board, connector CN9-right.

Pin	Port name	Usage area
1	PD7	Digital I/O, Interrupt, UART, SPI
2	PD6	Digital I/O, Interrupt, UART
3	PD5	Digital I/O, Interrupt, UART
4	PD4	Digital I/O, Interrupt, UART, SPI
5	PD3	Digital I/O, Interrupt, UART, SPI
6	GND	Ground voltage
7	PE2	Digital I/O, Interrupt, SPI
8	PE4	Digital I/O, Interrupt, SPI
9	PE5	Digital I/O, Interrupt, Timer, SPI
10	PE6	Digital I/O, Interrupt, Timer, SPI
11	PE3	Digital I/O, Interrupt
12	PF8	Digital I/O, Interrupt, Timer, UART, SPI
13	PF7	Digital I/O, Interrupt, Timer, UART, SPI
14	PF9	Digital I/O, Interrupt, Timer, UART, SPI
15	PG1	Digital I/O, Interrupt

Table A.7 Pin usage table for the STM32 board, connector CN10-left.

Pin	Port name	Usage area
1	AVDD	External analog voltage/voltage reference pin
2	AGND	Analog ground voltage
3	GND	Ground voltage
4	PB1	Digital I/O, Interrupt, Timer, ADC
5	PC2	Digital I/O, Interrupt, SPI, ADC
6	PF4	Digital I/O, Interrupt, ADC
7	PB6	Digital I/O, Interrupt, Timer, UART, I^{2}C
8	PB2	Digital I/O, Interrupt, SPI
9	GND	Ground voltage
10	PD13	Digital I/O, Interrupt, Timer, UART, I^{2}C
11	PD12	Digital I/O, Interrupt, Timer, UART, I^{2}C
12	PD11	Digital I/O, Interrupt, UART, I^{2}C
13	PE2	Digital I/O, Interrupt, SPI
14	GND	Ground voltage
15	PA0	Digital I/O, Interrupt, Timer, UART, ADC
16	PB0	Digital I/O, Interrupt, Timer, UART, ADC
17	PE0	Digital I/O, Interrupt, Timer, UART

Table A.8   Pin usage table for the STM32 board, connector CN10-right.

Pin	Port name	Usage area
1	PF13	Digital I/O, Interrupt, I^2C
2	PE9	Digital I/O, Interrupt, Timer, UART
3	PE11	Digital I/O, Interrupt, Timer, UART, SPI
4	PF14	Digital I/O, Interrupt, I^2C
5	PE13	Digital I/O, Interrupt, Timer, SPI
6	PF15	Digital I/O, Interrupt, I^2C
7	PG14	Digital I/O, Interrupt, Timer, UART, SPI
8	PG9	Digital I/O, Interrupt, UART, SPI
9	PE8	Digital I/O, Interrupt, Timer, UART
10	PE7	Digital I/O, Interrupt, Timer, UART
11	GND	Ground voltage
12	PE10	Digital I/O, Interrupt, Timer, UART
13	PE12	Digital I/O, Interrupt, Timer, SPI
14	PE14	Digital I/O, Interrupt, Timer, SPI
15	PE15	Digital I/O, Interrupt, Timer
16	PB10	Digital I/O, Interrupt, Timer, UART, SPI, I^2C
17	PB11	Digital I/O, Interrupt, Timer, UART, I^2C

Bibliography

Åström, J.K. and Murray, R.M. (2008). *Feedback Systems: An Introduction for Scientists and Engineers*. Princeton University Press.

Åström, J.K. and Wittenmark, B. (2013). *Adaptive Control*, 2e. Dover Publications.

Bertsekas, D.P. (2012). *Dynamic Programming and Optimal Control*, vol. 1, 4e. Athena Scientific.

Braunl, T. (2006). *Embedded Robotics Mobile Robot Design and Applications with Embedded Systems*, 2e. Springer-Verlag.

Burns, R.S. (2001). *Advanced Control Engineering*. Butterworth-Heinemann.

Chen, C.T. (2006). *Analog and Digital Control System Design: Transfer-Function, State-Space and Algebraic Methods*. Oxford University Press.

Chien, K.L., Hrones, J.A., and Reswick, J.B. (1952). On the automatic control of generalized passive systems. *Transactions of the ASME* 74: 175–185.

Cohen, G.H. and Coon, G.A. (1953). Theoretical consideration of retarded control. *Transactions of the ASME* 75: 827–834.

Corke, P. (2017). *Robotics, Vision and Control Fundamental Algorithms in MATLAB*, 2e. Springer.

Dorf, R.C. and Bishop, R.H. (2010). *Modern Control Systems*, 12e. Prentice Hall.

Fliess, M., Levine, J., Martin, P., and Rouchon, P. (1992). On differentially flat nonlinear systems. *IFAC Proceedings Volumes* 25: 159–163.

Forrai, A. (2013). *Embedded Control System Design A Model Based Approach*. Springer.

Franklin, G.F., Powell, J.D., and Workman, M.L. (2006). *Digital Control of Dynamic Systems*, 3e. Ellis-Kagle Press.

Ghosh, S. (2004). *Control Systems: Theory and Applications*. Pearson.

Golnaraghi, F. and Kuo, B.C. (2010). *Automatic Control Systems*, 9e. Wiley.

Goodwin, G.C., Graebe, S.F., and Salgado, M.E. (2000). *Control System Design*. Pearson.

Gopal, M. (2003). *Digital Control and State Variable Methods: Conventional and Neuro-Fuzzy Control Systems*, 2e. McGraw-Hill.

Embedded Digital Control with Microcontrollers: Implementation with C and Python,
First Edition. Cem Ünsalan, Duygun E. Barkana, and H. Deniz Gürhan.
© 2021 The Institute of Electrical and Electronics Engineers, Inc. Published 2021 by John Wiley & Sons, Inc.
Companion website: www.wiley.com/go/Unsalan/Embedded_Digital_Control_with_Microcontrollers

Hristu-Varsakelis, D. and Levine, W.S. (2005). *Handbook of Networked and Embedded Control Systems*. Birkhauser.

Ibrahim, D. (2006). *Microcontroller Based Applied Digital Control*. Wiley.

Jordan, D.W. and Smith, P. (2007). *Nonlinear Ordinary Differential Equations*, 4e. Oxford University Press.

Kalman, R.E. (1960). A new approach to linear filtering and prediction problems. *Journal of Basic Engineering* 82 (1): 35–45.

Kalman, R.E. (1963). Mathematical description of linear dynamical systems. *Journal of the Society for Industrial and Applied Mathematics, Series A: Control* 1 (2): 152–192.

Kuo, S.M. and Morgan, D.R. (1996). *Active Noise Control Systems: Algorithms and DSP Implementations*. Wiley-Interscience.

Ledin, J. (2004). *Embedded Control Systems in C/C++: An Introduction for Software Developers Using MATLAB*. CMP Books.

Leith, D.J. and Leithead, W.E. (2010). Survey of gain-scheduling analysis and design. *International Journal of Control* 73: 1001–1025.

Ljung, L. (1999). *System Identification: Theory for the User*, 2e. Prentice Hall.

Mandal, A.K. (2010). *Introduction to Control Engineering: Modeling, Analysis and Design*, 2e. New Age International Pvt. Ltd.

Moudgalya, K.M. (2008). *Digital Control*. Wiley.

Ogata, K. (1995). *Discrete-Time Control Systems*, 2e. Pearson.

Ogata, K. (2009). *Modern Control Engineering*, 5e. Pearson.

Oppenheim, A.W. and Schafer, R.W. (2009). *Discrete-Time Signal Processing*, 3e. Prentice Hall.

Oppenheim, A.W., Willsky, A.S., and Nawab, S.H. (1997). *Signals and Systems*, 2e. Prentice Hall.

Palm, W. (2013). *System Dynamics*, 3e. McGraw-Hill.

Phillips, C.L., Nagle, H.T., and Chakrabortty, A. (2015). *Digital Control System Analysis and Design*, 4e. Pearson.

Pintelon, R. and Schoukens, J. (2001). *System Identification. A Frequency Domain Approach*. Wiley-IEEE Press.

Proakis, J.G. and Manolakis, D.K. (1995). *Digital Signal Processing: Principles, Algorithms and Applications*, 3e. Prentice Hall.

Skogestad, S. and Postlethwaite, I. (1996). *Multivariable Feedback Control: Analysis and Design*. Wiley.

Söderström, T. and Stoica, P. (1989). *System Identification*. Prentice Hall.

Starr, G.P. (2006). *Introduction to Applied Digital Control*. Draft.

STMicroelectronics (2015). *UM1925 User Manual*, docid028121 rev 1 edition.

Tewari, A. (2002). *Modern Control Design with MATLAB and Simulink*. Wiley.

Ünsalan, C. and Gürhan, H.D. (2013). *Programmable Microcontrollers with Applications: MSP430 LaunchPad with CCS and Grace*. McGraw-Hill.

Ünsalan, C., Gürhan, H.D., and Yücel, M.E. (2017). *Programmable Microcontrollers: Applications on the MSP432 LaunchPad*. McGraw-Hill.

Ünsalan, C., Yücel, M.E., and Gürhan, H.D. (2018). *Digital Signal Processing using Arm Cortex-M based Microcontrollers: Theory and Practice*. Arm Education Media.

Wescott, T. (2006). *Applied Control Theory for Embedded Systems*. Newnes.

Xue, D., Chen, Y.C., and Atherton, D.P. (2007). *Linear Feedback Control Analysis and Design with MATLAB*. SIAM.

Yiu, J. (2013). *The Definitive Guide to ARM® Cortex®-M3 and Cortex®-M4 Processors*, 3e. Newness.

Yu, W., Wen, G., Chen, G., and Cao, J. (2016). *Distributed Cooperative Control of Multiagent Systems*. Wiley.

Zhou, K. and Doyle, J.C. (1997). *Essentials of Robust Control*. Pearson.

Ziegler, J.G. and Nichols, N.B. (1942). Optimum settings for automatic controllers. *Transactions of the ASME* 64: 759–768.

Index

Embedded Digital Control with Microcontrollers: Implementation with C and Python,
First Edition. Cem Ünsalan, Duygun E. Barkana, and H. Deniz Gürhan.
© 2021 The Institute of Electrical and Electronics Engineers, Inc. Published 2021 by John Wiley & Sons, Inc.
Companion website: www.wiley.com/go/Unsalan/Embedded_Digital_Control_with_Microcontrollers